GEOMORPHOLOGY SANS FRONTIÈRES

INTERNATIONAL ASSOCIATION OF GEOMORPHOLOGISTS

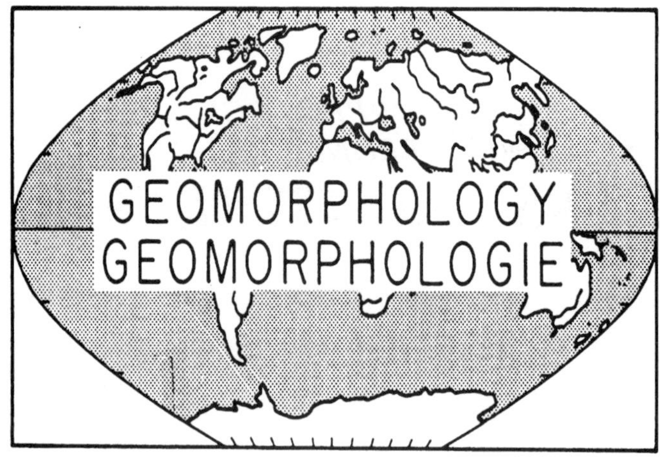

Publication No. 1

THE EVOLUTION OF GEOMORPHOLOGY
A Nation-by-Nation Summary of Development

Edited by H.J. Walker *and* W.E. Grabau

Publication No. 2

RIVER GEOMORPHOLOGY

Edited by Edward J. Hickin

Publication No. 3

STEEPLAND GEOMORPHOLOGY

Edited by Olav Slaymaker

Publication No. 4

GEOMORPHIC HAZARDS

Edited by Olav Slaymaker

Publication No. 5

LANDSLIDE RECOGNITION

Edited by Richard Dikau, Denys Brunsden, Lothar Schrott *and* Maïa-Laura Ibsen

Publication No. 6

GEOMORPHOLOGY SANS FRONTIÈRES

Edited by S. Brian McCann *and* Derek C. Ford

GEOMORPHOLOGY SANS FRONTIÈRES

Edited by

S. BRIAN McCANN
and
DEREK C. FORD
McMaster University, Canada

JOHN WILEY & SONS
Chichester · New York · Brisbane · Toronto · Singapore

National 01234 779777
International (+44) 1243 779777
e-mail (for orders and customer service enquiries): cs-books@wiley.co.uk
Visit our Home Page on http://www.wiley.co.uk
or http://www.wiley.com

Other Wiley Editorial Offices

John Wiley & Sons, Inc., 605 Third Avenue,
New York, NY 10158-0012, USA

Jacaranda Wiley Ltd, 33 Park Road, Milton,
Queensland 4064, Australia

John Wiley & Sons (Canada) Ltd, 22 Worcester Road,
Rexdale, Ontario M9W 1L1, Canada

John Wiley & Sons (Asia) Pte ltd, 2 Clementi Loop #02-01,
Jin Xing Distripark, Singapore 0512

Library of Congress Cataloging-in-Publication Data

Geomorphology sans frontières / edited by S.B. McCann & D.C. Ford.
p. cm. — (Publication : no. 6)
Papers from the Third International Geomorphological Conference held at McMaster University, Hamilton, Canada, Aug. 1993.
Includes bibliographical references and index.
ISBN 0-471-96600-2
1. Geomorphology — Congresses. I. McCann, S. Brian (Samuel Brian) II. Ford, Derek (Derek C.) III. International Geomorphological Conference (3rd : 1993 : McMaster University) IV. Series: Publication (International Association of Geomorphologists) : no. 6.
GB400.2.G448 1996 96–1961
551.4′1 — dc20 CIP

British Library Cataloguing in Publication Data

A catalogue record for this book is available from the British Library

ISBN 0-471-96600-2

Typeset in 10/12pt Times from editor's disk by MHL Typesetting Ltd, Coventry
Printed and bound in Great Britain by Bookcraft (Bath) Ltd, Midsomer Norton, Somerset
This book is printed on acid-free paper responsibly manufactured from sustainable forestation, for which at least two trees are planted for each one used for paper production.

Contents

Contributors

Frank Ahnert Geographisches Institut der RWTH, Templergraben 55, 52056 Aachen, Germany

Victor R. Baker Department of Geosciences, University of Arizona, Tuczon, Arizona, AZ 85721, USA

Hanna Bremer Geographisches Institut, Universität zu Koln, Albertus-Magnus-Platz, D-5000 Koln 1, Germany

Denys Brunsden Department of Geography, King's College, University of London, London WC2R 2LS, UK

Derek C. Ford Department of Geography, McMaster University, Hamilton, Ontario L8S 4K1, Canada

S. Brian McCann Department of Geography, McMaster University, Hamilton, Ontario L8S 4K1, Canada

Yoto Ota Department of Geography, Sen shu University, Kawasaki, 214-80, Japan

Anders Rapp Department of Physical Geography, University of Lund, Sölvegatan 13, S-22362 Lund, Sweden

John Shaw Department of Geography, University of Alberta, Edmonton, Alberta T6G 2H4, Canada

Olav Slaymaker Department of Geography, University of British Columbia, Vancouver, British Columbia V6T 1W5, Canada

Jean L.F. Tricart Centre de geographie appliquée, 3 Rue de L'Argonne, Strasbourg-Cedex 67083, France

C.R. Twidale Department of Geology and Geophysics, University of Adelaide, Adelaide, South Australia 5005, Australia

Eiju Yatsu 91 Iwase, Matsudoshi 271, Japan

Series Preface

This volume is the latest of four books emanating from the Third International Geomorphology Conference, held at McMaster University, Hamilton, Ontario, in August of 1993. All the papers in these volumes were independently reviewed before being accepted for publication. I regret, however, that we were unable to publish some excellent papers from the conference that were, unfortunately, on topics other than those covered in this series. I thank the editors Ted Hickin (*River Geomorphology*), Olav Slaymaker (*Steepland Geomorphology* and *Geomorphic Hazards*) and Brian McCann and Derek Ford (*Geomorphology Sans Frontières*) for their efficiency and diligence, the many reviewers of the papers, contributors, who showed great patience waiting to see their papers in print, and the staff at John Wiley, who have been most helpful and encouraging. I also thank Denys Brunsden, former president of the International Association of Geomorphologists, who provided the initial impetus for this series and continued support.

A.S. Trenhaile, Series Editor
University of Windsor, Ontario, Canada

Preface

The natural landscapes of the earth display a variety and complexity that is not matched on the other planetary bodies that are known. Modern geomorphologists, students of these landscapes, also display a wealth of different perspectives and approaches to their science. Reconstructions of the landscape history of regions using Davisian or climatogenic or dynamic equilibrium concepts continue to form a major part of the published literature. The plate tectonic revolution in geology has re-focused interest on the role of tectonic deformation over many differing time spans. The belated recognition during the 1970s of the quite high frequency of meteorite impact craters on earth's continents and of mega-flood relict landscapes on Mars, has opened new debate upon catastrophism. At the same time that this range of broad issues has been addressed, perhaps a majority of practising geomorphologists turned to the measurement of processes in the field in greater detail than ever attempted before; many also resorted to hardware modelling using flumes etc., in laboratories or to controlled experiments with rain machines and other devices in the field. Mathematical modelling of form evolution made important advances in its formalism in the 1960s and 1970s. The advent of inexpensive, convenient and very powerful computing in the 1980s then greatly expanded the scope for iterative modelling, which seems set to be a major growth area in the coming years. Fractal landform geometries, non-linear dynamics in the operation of geomorphic processes, and chaotic structure in natural systems are being investigated vigorously. Geomorphologists of all perspectives have applied their skills to the practical problems of protecting and managing the natural environment.

All of these approaches were on show at the Third International Geomorphology Conference and are represented in this book, which collects the plenary addresses and debates in one volume. The Conference took place at McMaster University, Hamilton, Canada, in August 1993. Previous conferences had been held in Manchester, England, in 1985 and in Frankfurt, Germany, in 1989, but the Hamilton conference was the first to be held under the auspices of the International Association of Geomorphologists, which was formed at the end of the Frankfurt meeting. The Hamilton conference was attended by over 650 delegates from 41 nations, who contributed nearly 800 abstracts of papers which were presented in a variety of formats – half-day symposia, paper sessions and poster sessions. As at other large meetings, concurrent sessions were unavoidable; at times delegates had a choice of ten simultaneous papers.

However, within this very busy and diverse schedule there were a number of occasions when there was a single focus of attention for all delegates. These were the pairs of plenary lectures, representing international and Canadian research, which occupied the middle two hours of four of the conference days, the debate concerning the ten principles of geomorphology propounded by Denys Brunsden at Frankfurt, and Denys' closing academic address in Hamilton, as retiring president of the association. We have adopted the title of this address — Geomorphologie Sans Frontières — as the title for this volume, which contains the text of five of the plenary lectures, an account of the debate, with the opening statements of the panel of five discussants, and the presidential address itself.

The presidential address (Chapter 1) is essentially a continuation of Denys' statement about the nature and development of geomorphology made four years previously. The first address offered ten propositions as a framework for the development of the subject: the second address, published here, suggests ten responsibilities for geomorphologists and the new International Association. The range of opinions expressed by the five discussants of the first paper (Hanna Bremer, C.R Twidale, Eiju Yatsu, Victor Baker and Jean Tricart), which are published here (Chapter 2) with a brief introduction by Olav Slaymaker, suggests that the second paper may also serve to generate discussion, argument and exchange of ideas.

The plenary lecturers represent different fields of geomorphology and each has made a major contribution to the subject. They were invited by the Programme Committee to address a topic of their own choosing for an hour and to contribute a written text for publication. They were not asked to adopt or espouse a particular methodological or philosophical position, nor were they constrained to consider a particular set of phenomena. Not surprisingly, they chose topics near to their hearts, reviewing findings and problems which had occupied much of their working lives as geomorphologists. In reading the papers it is possible to discern something of the different national styles of geomorphological enquiry but the overwhelming impression is the stamp of the individual authors' personalities.

The three international plenary papers in this volume deal with widely divergent topics. Frank Ahnert (Chapter 3) provides a timely examination of the role of models in geomorphology, focusing on process response models, with illustrations from his own work over 30 years. The development of cuestas and structural benches on layered rocks, and inselbergs in lithologically and structurally homogenous rocks are two examples. Yoko Ota (Chapter 4) discusses the implications of coral reef studies, using radiometric dating of fossil corals in reef terraces, for reconstructing regional patterns and rates of Late Quatemary tectonic deformation. She instances the contrasting styles and rates of deformation of coral reef terraces on islands on the overriding Eurasia Plate with similar features developed on islands on the subducting Philippine Sea Plate. Anders Rapp (Chapter 5) considers the evidence which indicates that blockfields and other surface periglacial phenomena may be present under cold-based glacier ice in mountain regions.

The two Canadian plenary papers, like Rapp's, deal with cold-region geomorphology, not surprisingly, perhaps, given the overwhelmingly glacial and periglacial character of our landscapes. Derek Ford's contribution (Chapter 6), however, concerns the karst geomorphic system, the underground part of which is not only preserved below surface glacial conditions but may develop apace. He provides examples of the wide range of phenomena exhibited by karst in a cold climate. John Shaw's paper (Chapter 7) is the longest in the book. It is a very thorough and careful presentation, with supporting evidence, of the hypothesis that meltwater processes were of major importance in the development of deglacial landscapes as Pleistocene ice sheets decayed. In particular, he focuses on the decay of the Laurentide ice sheet and related outburst floods.

S. Brian McCann and Derek C. Ford
Hamilton, May 1995

1

Geomorphologie Sans Frontières

Presidential Address, 1993 International Association of Geomorphologists, McMaster University, Canada

DENYS BRUNSDEN

Department of Geography, King's College, London, UK

Inside my closed eyes
I see veins of leaves
I see the colour of the earth
I see the burning sun
Hot, red and molten
And its burning burns and my flesh turns to run away

Inside my closed ears
I hear tumbling sea
I see the colour of the earth
I see bursting clouds
Meandering, tumbling and flowing
And my tears burst and my body thirsts to run away
But the wind pushes me back
And the water carries me
And the earth runs through my fingers
And the fire bursts out of my soul
We will reach that goal
We will unfold a better world.

(Elemental Birthday, Song cycle by Toyah Wilcox and Sammy Hurden, Earth 1992)

INTRODUCTION

The intention of this lecture is to discuss the responsibilities of our subject, association and profession.

As I sat down to write the script, the joyful news came that Elizabeth and I had become grandparents for the third and fourth time, James David in England and Yi Ting Lin in Taiwan.

Geomorphology Sans Frontières. Edited by S. Brian McCann and Derek C. Ford.

At the same time the English newspapers were full of photographs of children in Sudan, which were designed to shock. The contrast in the well-being and opportunity of the children was obvious and painful, and made me question just what the word "responsibility" meant.

These coincidences were too powerful for me to ignore and this address therefore also became a personal exploration of the meaning of "being a geomorphologist in the modern world". It is also the first Presidential Address to a fledgling discipline on the international stage. It is therefore appropriate to make an examination, on your behalf, of the nature of geomorphology as a profession. Do we, for example, have a special role to play in the wise management of earth?

The present political and socio-economic world context for the subject is the disparity between the wealth of nations, a rapid increase in the number of people living at subsistence levels, and a determination that the standard of living of the people of the developing countries should be raised.

The questions we are asked by the relevant world agencies concern whether we can assist in the efficient discovery, assessment and wise use of the earth's finite resources, to help prevent environmental deterioration and to mitigate the effect of natural hazards?

The questions we might ask of ourselves are whether we have a desire and capability to apply our techniques, methods of analysis and conceptual understanding of the behaviour of earth surface materials, processes and landforms to these problems? The question I might ask as your President is this: If we do wish to contribute, what must we do as a profession to achieve this aim?

SHOULD WE CONTRIBUTE?

At the time that James David was born, I was reading "The reenchantment of geomorphology" by Baker and Twidale (1991) and had just come to a discussion of the importance to modern scientific synthesis, of global problems, such as greenhouse warming, ozone depletion, deforestation and desertification. Two phrases stood out, "Geomorphology", they said, "might well assert itself as an integrative expression of what it is important to know of nature" and that one of the things we need to know is "that which we may act upon without fear". (W. Clifford in Bronowski 1979).

Baker and Twidale (1991), following Jennings (1973), also warned that too close an adherence to any particular geo-ideological bandwagon might "stifle creativity, insight, and intellectual satisfaction". They consider that neither the scientific revolutions which have affected both geology and geography in the last 40 years (plate tectonics, scientific method), nor the adoption of successive paradigms (landscape evolution, climatic geomorphology, process studies) or doctrines (antihistoricism, theoreticism, utilitarianism, realism) have provided a satisfactory unifying theme for the subject. They believe that applied work depends on the continuous generation of fundamental ideas and state that, "Application cannot be the driving purpose of Geomorphology even though geomorphologists have much of value to contribute in this area."

In consequence, Baker and Twidale go on to appeal for a return to "a time of discovery" in which geomorphologists will "again find magic, inspiration and messages in the landscape" ... "integral to the study of a dynamically vibrant planet". "We look forward to the time" they wrote, "when geomorphologists, armed with physical theory, not stifled with abstractions, but bent on discovery, will experience the excitement of unraveling the development of landscapes through time, of conversing with the landscape and responding to its promptings".

Actually, I feel that we have always done this and find it strange that two eminent geomorphologists feel the need to express such a sentiment, but it cannot be denied that other geomorphologists also believe that we should be very wary of the applied arena. Chorley (1978, p.10) drew attention to the growth of applied geomorphology as follows:

> First, it is clear that, for reasons of syllabus, popularity (professional, student and lay), finance, "relevance'' and the like, geomorphology is becoming increasingly committed, if not wholly then in significant part, to a role involving its relationship to human well-being and aspirations.

Supported by Thornes (1978) he indicated that there was a gulf between the theoretical bases made available to us through scientific method and those required by utilitarian and social doctrinaire theories and that that gulf divided radically different environmental aims. He also showed that, because the growth of theory is related to the intellectual climate, the theory and the work we do is vulnerable to the pressures of convention.

As Walker (1978) said in the same conference:

> as applied research is aimed at solving specific problems, usually those with pressing economic significance, it is easily and often emphasised at the expense of basic research. Such emphasis will almost certainly lead to fewer fundamental discoveries, discoveries that actually make applied research meaningful.

The implication is that unless we are aware how our theory may be affected by the conditions under which it is developed, then applied research may prove to be a bad thing in the long term. The questions to be asked are whether either the viewpoint or the implication are true.

My own view is that this is only acceptable if we are acting on a stage unfettered by a sense of responsibility to those whose "conversations with earth'' (Cloos 1953) are determined by survival rather than scientific freedom.

The realities of our world are that many people, some of them our geomorphological colleagues, are not free but operate within significant constraints. For some the frontiers to be crossed are merely physical but for others they are economic, social, political ethno-territorial, military and organizational. Presidents of international associations become painfully aware of these barriers as they receive requests for guidance, financial help, travel grants, information, equal opportunity and justice. These are understandable from those who wish for nothing more than opportunity or to learn and we all respond by seeking funds to facilitate mobility or information dissemination. Our hosts here in Canada have made noble efforts in this way and many would not have attended the Hamilton conference without that generosity.

The real problem comes, however, when the frontier is one of hopelessness and despair. As we cross that frontier, scientific freedom becomes a meaningless concept. Those of us who have academic freedom also have unavoidable responsibilities to those who do not. To achieve this we must contribute to the practical arena of our subject and to recognize that freedom is a luxury that many simply cannot afford.

It is my contention the Association must steadily improve its efforts and resources to work both for the exchange of scientific information on an international basis and for the general well-being of the earth and its inhabitants.

In saying this, of course, it should entirely adopt and defend the view that a civilized society should have the freedom to study without fear of oppression or directives.

This may be expressed as two primary responsibilities:

Responsibility l. Equal Opportunity
To work for the equal opportunity of all people to cross the frontiers of knowledge.

Responsibility 2. Scientific Freedom
To support the ideal of the scientific freedom to study without the pressures of convention, politics or oppression.

THE NATURE OF APPLIED GEOMORPHOLOGY

The subject I address here is what happens when, with that freedom, we choose to study with a practical applied purpose in order to satisfy the first responsibility. Can that yield the sort of research which leads us to a reenchantment with our subject? Is such an approach bad for the subject?

These questions can only be answered after we have examined what is meant by the term "applied geomorphology". A suitable definition is "the application of geomorphological techniques and analysis to a planning, conservation, resource evaluation, engineering or environmental problem." (Brunsden et al. 1978).

Over recent years it has become apparent that geomorphological studies are capable of providing information of value to anyone who is required to operate in the natural environment (Figure 1.1). Most geomorphologists (e.g. Cooke 1987) believe that all parts of their subject are applicable to human problems because we are always dealing with the forms and processes which are used by humans and which have to be managed sensitively if we are to avoid difficulties such as land degradation or natural hazard.

Applied geomorphology can be divided into three types of work.

1. Problem identification and baseline surveys

This involves desk studies, walkover surveys, reconnaissance, routine resource assessments or standard environmental statements which, one hopes, are carried out before a project begins. It is the most important aspect of applied work because site investigation and design can only proceed efficiently if the problems are recognized. This may seem self-evident, but we are not facing a leisurely academic research situation. Time is always short, there are tight time schedules because there are contractual deadlines, there is a need to be right because large sums of money and perhaps lives may depend on the operational decisions. The dilemma is that data are often few and far between or of very low quality and yet there is little time to establish well-controlled data collection systems (Cooke and Doornkamp 1992).

At this point judgements have to be made on the nature, origin, age and relationships of the landforms. Deposits must be assessed and related to the environments in which they were created. Processes have to be estimated from fragmentary records, narrative, print,

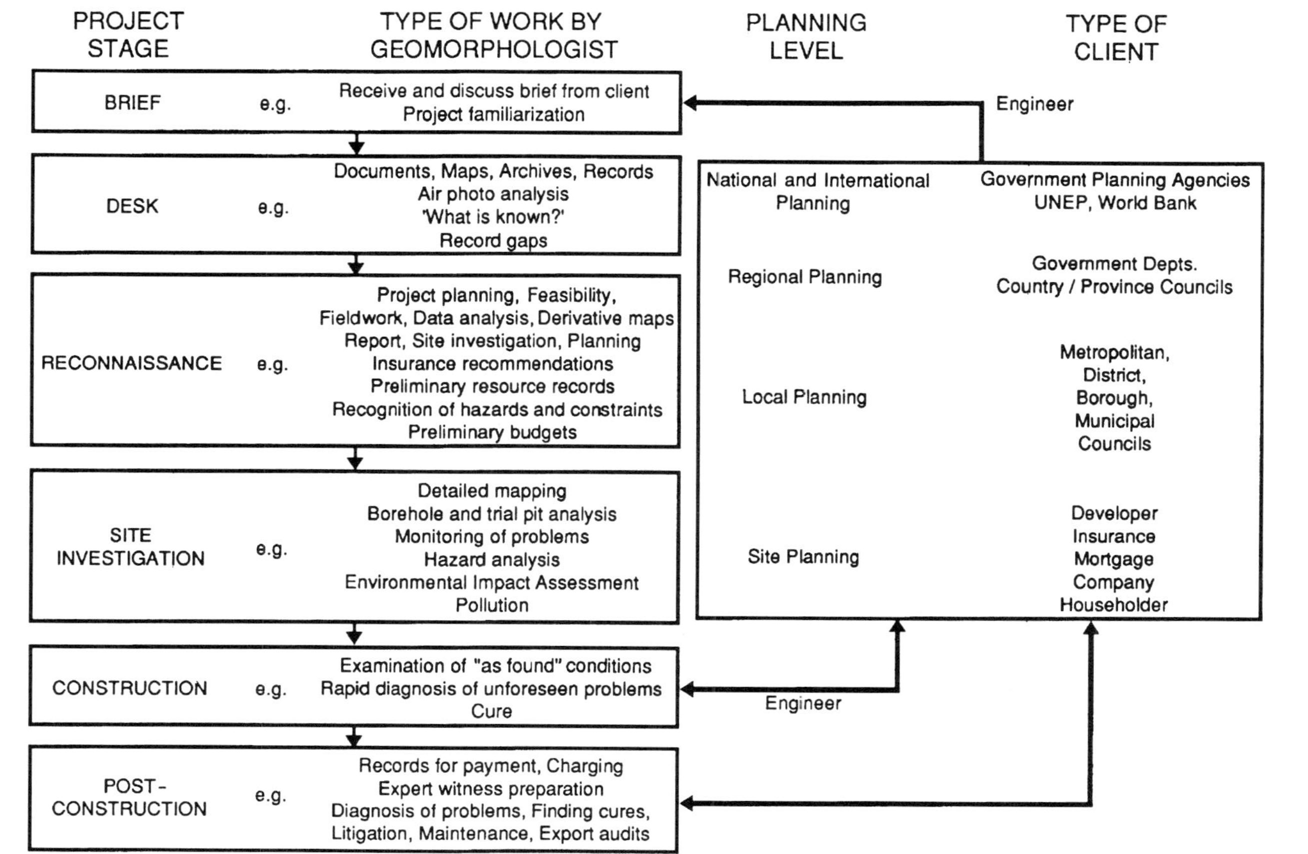

FIGURE 1.1 The nature of applied geomorphology

photograph, map and other historical archive materials. There is a demand for that unique skill of being able to estimate process from the evidence of the landforms. Under these circumstances there is no substitute for an ability to "read the ground". We also have recourse to the strategies of translating theoretical, laboratory or computer modelling to the site scale or of using space–time, analogue and historical comparisons to evaluate the possible responses (Cooke and Doornkamp 1992).

The aim of this type of interdisciplinary work is to give the most accurate advice possible backed up by a statement of the information used and the controlling constraints. There is no substitute for experience, intuition and a willingness to give an opinion, fully qualified by an honest statement of the criteria used for the judgement. The prospect of liability becomes a powerful motive for good science!

This requires a broad training in geomorphology, geology, pedology, hydrology and other earth sciences as well as a good spatial perception of the relationships involved. This incidentally, is the main reason why the IAG did not seek independent Scientific Union status under ICSU but are delighted with our Affiliate status to the IGU and IUGS. To build subject bridges is a *raison d'être* for an international association and we are actively involved with supporting the work of the existing working groups and commissions.

2. Innovative research

The second type of work which can be carried out at any stage of a project, but usually during site investigation, covers innovative studies which are necessary because an applied problem is recognized, where a disaster has occurred in an unknown area or situation, and/or where there are no current standard practices, techniques of measurement, recording or analysis. Usually the situations are serious and there is more time and finance to allow expensive investigations, University contract research often benefits at this point and there may be significant spin-offs in fundamental scientific discovery and equipment development. An example which has become a model for other arid areas is the development of salt hazard assessment techniques by the Bahrain Surface Materials Resources Survey to act as a planning guideline in areas of the capillary rise of saline water into building foundations (Brunsden et al. 1979; Doornkamp et al. 1980; Cooke et al. 1982) (Figure 1.2). Indeed a case will be made later that it is such work that has led to many of the most important advances of our subject in the last 50 years.

Despite this it has to be admitted that 90% of the work will be routine. It is a lucky scientist who is invited to carry out research which will establish new standards for the profession. As we shall see this has important implications for the geomorphologist acting as technician on standards and codes of practice.

3. Advisory (including insurance and litigation)

The third type of applied work concerns the geomorphologist acting as an advisory scientist or expert witness in an insurance or litigation situation. Insurance work will usually concern assessments of viability or risk a type of work which seems to be increasing in volume. Litigation work will involve tests of breach of contract, negligence or, if a public authority is involved, a breach of statutory duty (Reynolds and King 1988). The former usually takes place before a project begins, hopefully at the planning phase.

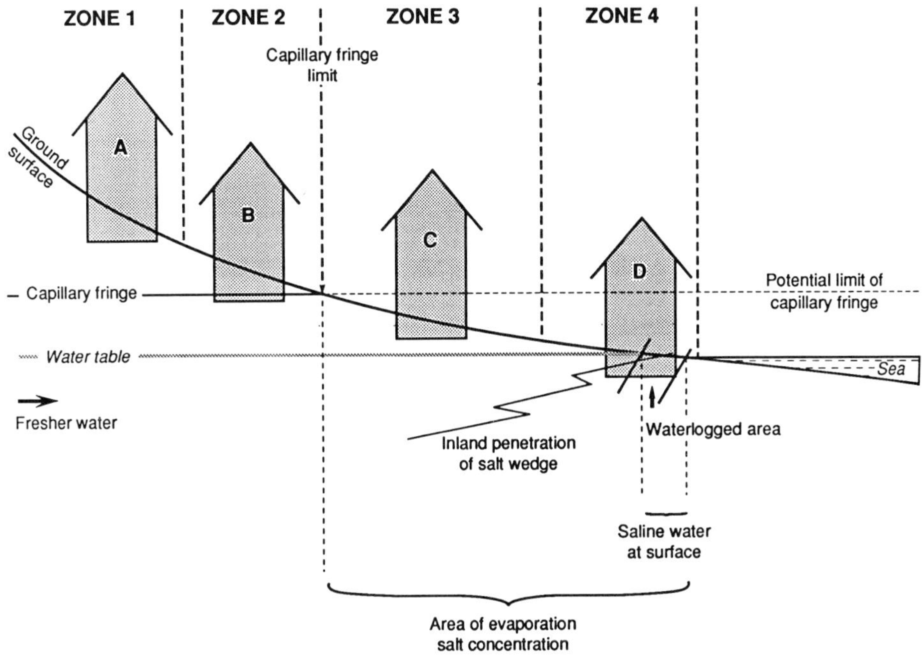

ZONE 1 Building foundations **(A)** do not reach capillary fringe.

ZONE 2 Building foundations **(B)** do reach capillary fringe.

ZONE 3 Building foundations and walls **(C)** affected by potential limit of capillary fringe and surface salt concentration.

ZONE 4 Building foundations and walls **(D)** affected by one or more of inland sea-water penetration, waterlogging, salt concentration and potential limit of capillary fringe.

FIGURE 1.2 The salt hazard mapping model developed by the Bahrain Surface Materials Resources Survey (Doornkamp et al. 1980)

A geomorphologist contemplating such work should consider the task with care. The role is very demanding and can be time consuming in a way never before experienced. In litigation the expert will be asked to give initial advice during which the possible contribution will be thoroughly assessed. This can be a very searching procedure. If it is decided to go ahead there will be a lengthy period during which preparations are made for trial. This period will involve a lot of reading and research, some of it intellectually demanding, some necessary but very tedious. First-time witnesses should discover just how big a commitment is required before agreeing to a job!

After the report-preparation stage the witness will be involved in the presentation of evidence, examination in chief, cross-examination and re-examination stages. It will be necessary to listen to the evidence of other witnesses and to advise the Counsels on the scientific significance of the statements made. It may be necessary to take part in the settlement negotiations if further explanations are required. Throughout these proceedings it is necessary to give an honest, clear view and to prepare thoroughly. The expert must not be biased or prejudiced and should always adopt a reasonable view. There is a responsibility to do a well-prepared job for the client but in the end the responsibility is to the Court which is seeking to establish the truth of the case.

THE STAGES OF APPLIED WORK

Within the types of work carried out by the applied geomorphologist will be small-scale mapping including land systems evaluation and other baseline surveys; large-scale surveys, at a scale of 1:250–1:10 000, which will involve the detailed study of areas, routes and specific sites; process monitoring; historical evaluations and appreciation of the time dimension; hardware and numerical modelling. The work will be carried out at four stages (Brunsden et al. 1975a):

1. *Reconnaissance stage*
At this stage work takes place wherever there is a need for a rapid assessment of ground conditions, steepness, drainage, resources, hazard etc. This stage is used for problem identification, avoidance of recognized problems (e.g. preliminary road alignment) and for the design of the site investigation.

2. *Site investigation stage*
In site investigation, work is directed at a chosen area, route or site and carried out parallel to site investigations by engineering geologists, soil mechanics specialists, foundation engineers, landscape architects, conservationists etc. (Figure 1.3). The aim is to fully understand the ground conditions and processes so that detailed designs can be made, refined and improved. Alignment, slope, foundation and drainage designs as well as land-take requirements and environmental impacts are all decided at this time.

3. *Construction stage*
It is to be hoped a geomorphologist will not be required at this time because that would mean that something had gone wrong or an unforeseen problem has emerged late in the project. Unfortunately, earth scientists of all kinds have found themselves employed in this way, occasionally because a new problem emerged but more often because they had not been used in the first place to contribute to the design.

Work at this stage will involve diagnosis for a cure and mapping for the record, to provide information to be used in future work or as the basis for payment. It will also be necessary to prepare accurate records as evidence for contractual, negligence or third-party claims. These usually involve cases where there is damage to neighbouring land and property, construction delays or payment for the discovery of ground conditions which were different to those that were predicted.

4. *Post-construction*
After a project has been completed there may still be the need to monitor the performance of a design and to carry out post-audit assessments. This is an undeveloped area because few projects include a budget for this purpose. Normally the work involves re-assessments

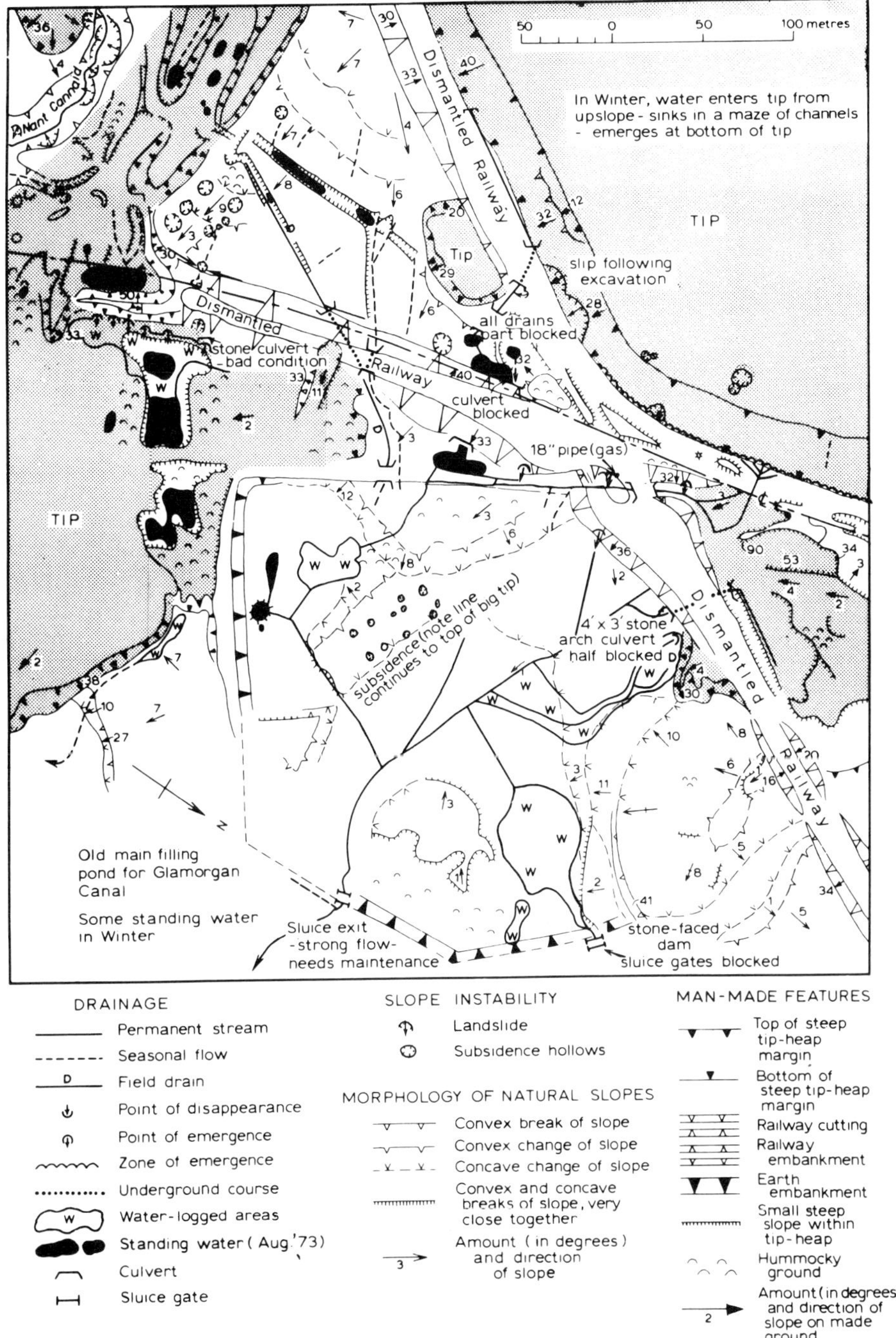

FIGURE 1.3 An early site investigation map made for Rendel Palmer and Tritton, Civil Engineers, in 1975 by D. Brunsden, J.C. Doornkamp and D.K.C. Jones when trying to prove the value of geomorphology to engineering. RPT went on to become a champion of the subject. (Brunsden et al. 1975a)

only if the design has not worked or if there is to be an upgrading. In a few cases, such as the innovative action of the Frontier Works Organisation of the Karakoram Highway in Pakistan, geomorphological assessments are included in the regular maintenance programme because there are unavoidable geomorphological hazards (Figure 1.4).

Lastly, the individual needs of many different clients must be appreciated. Services will be provided for:

Individuals — who want a cheap, reliable service which gives commonsense advice and, often, reassurance about local hazards or developments.

Developers — who need the best possible independent advice on the viability of a proposed scheme in relation to ground conditions, processes and impacts on neighbours or environments.

Consulting engineers — faced with active natural processes such as ground stability, erosion, the behaviour of river systems or the complex behaviour of the coastline. The engineer is also likely to be at the sharp end of the development projects needed to deal with the human problems with which we began this address.

Planners–local government — by making an assessment of the ground conditions, diagnosing problems and providing policy management and development strategies.

Planners–government ministeries — by providing policy advice concerning the management of the environment and a rapid response for the assessment of natural disasters. This is particularly important in the 1990s, the International Decade for Natural Disaster Reduction.

Reinsurance officers — who require an assessment of ground and environment conditions as they respond to a request for cover or to a claim for damage or loss.

Lawyers — who need expert witnesses or other litigation guidance.

International agencies — who require all forms of help, especially environmentally sensitive approaches to development; evaluations of the viability of schemes; expert panels to oversee or approve major developments or expenditures; rapid response or report teams following a natural disaster; technical workshops and educational programmes.

TRANS-FRONTIER GEOMORPHOLOGICAL ISSUES

One of the most interesting phenomena of our time is the strong emergence of trans-frontier environmental, resource, hazard and resource issues. These often accompany or cause ethno-territorial tensions, border incidents, conflicts and refugee movements.

In many cases the solutions can involve geomorphological help or information. As geomorphologists we recognize that a perennial but dynamic feature of the earth's surface is that human frontiers rarely coincide with the boundaries of natural systems. It is also sadly true that despite huge lessons from the past, such as the Tennessee Valley Authority and the dustbowl solution, we have still not learnt that geomorphological problems arise when human action in one part of a system is taken without consideration of its effects on other components. Nor, in many cases, do we have knowledge of the side-effects. Wherever there are strong interlinkages in the systems there is a potential for change and degradation.

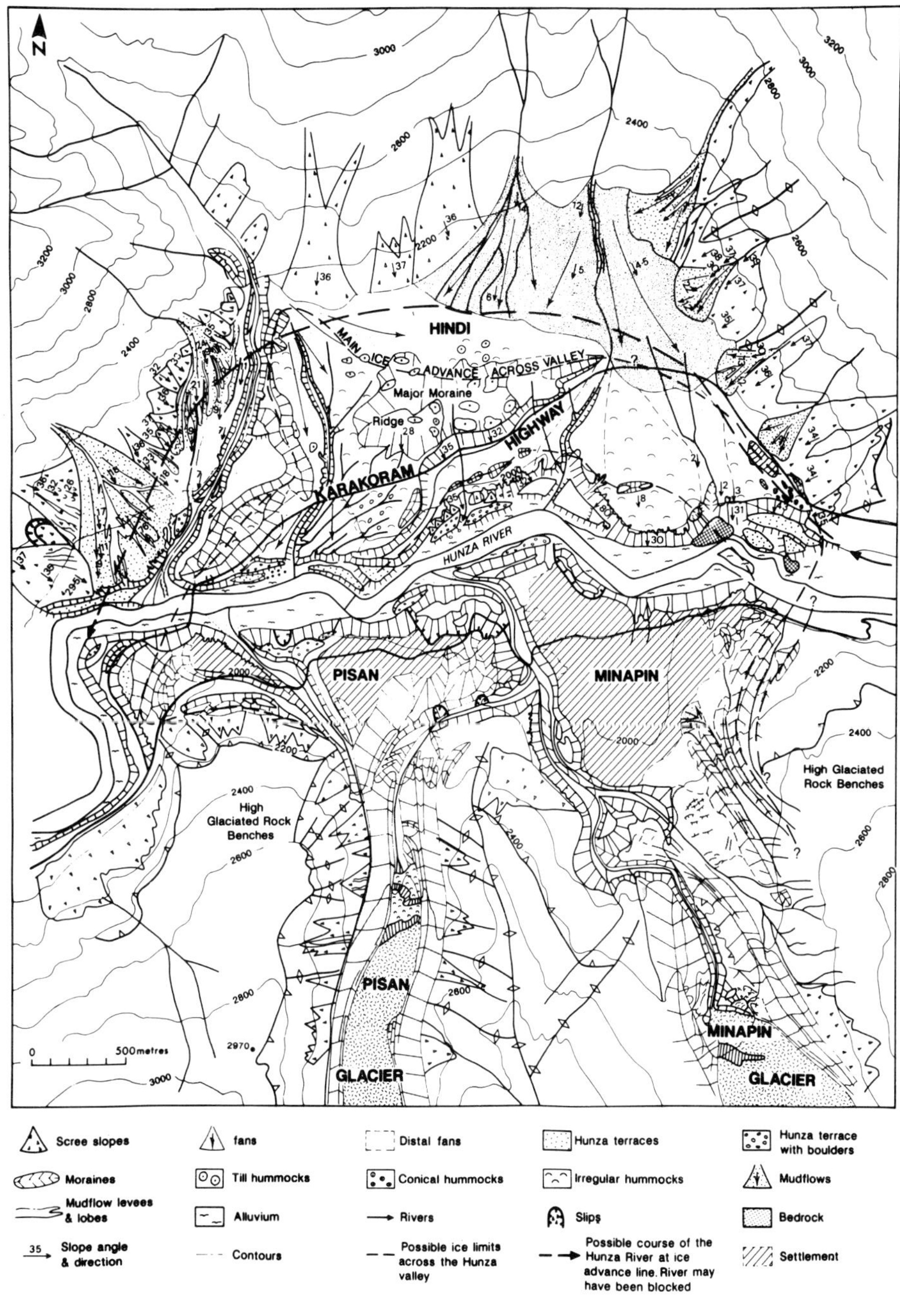

FIGURE 1.4 Typical map produced for post-construction geomorphological assessments, the Karakoram Highway (Goudie et al. 1984)

Nor should this be thought of as concerning big areas only. Similar issues occur at a local scale. The creation of harbours or groins on one piece of a coast and the cause of erosion on another, usually in someone else's backyard, is the classic problem.

On an international scale the problems range over the use of unoccupied territory, such as Antarctica for scientific research or potential resources; the scramble for space and resources in the international oceans as the new technologies unlock the ocean floor; the erosion and pollution problems of sea and land areas that are shared between nations and have common geomorphological system interactions. The Black Sea, the Mediterranean, the South China Sea or the Bay of Bengal might serve as examples and many delta areas of the world experience similar issues. Perhaps even more difficult is the export of dust from arid areas. The wind knows no human frontiers.

All of these issues raise the problem of how to develop an understanding of the criteria which might be used to share resources and mitigate common hazards. It is a valid (but perhaps naive) point that the scientific view might reach consensus where national interests fail!

Nowhere is this more important or environmentally necessary than in the upstream–downstream situation of the great international drainage basins. An incomplete list includes:

— Egypt's concern that use of water in the upper Nile catchment will affect agriculture and water availability in the Aswan Dam and the delta.
— The claims over Jordan water and the aquifers by the inhabitants of the West Bank, Israel, Syria and Jordan.
— The choices to be made between Turkey, Iraq, Iran and Syria over the use of the Tigris–Euphrates. For example the South East Anatolia project with a proposal for 22 large dams, 17 hydro-electric stations and thousands of square kilometres of irrigated land, including the Ataturk Dam, will have an unknown effect on siltation and downstream incision as far as the delta at the head of the Gulf where drainage of the marshes is already well advanced.
— The use of the Rio Grande between the USA and Mexico.
— The management of the Zambesi between its eight bordering countries.
— The use of the Mekong.
— The possible effect of land use in Nepal on river behaviour in the lowlands of India or Bangladesh.

Hungarian newspapers (see *The Budapest Sun* **10**, May 6–12 1993) have recently been full of discussion about the multiple use of the Danube, in this case an informed debate about the possible effects of the Gabcikovo Dam and hydro-electric plant in Slovakia and the possible Dunakiliti reservoir in Hungary on the groundwater levels in the downstream alluvial fans at Szigetkoz in Hungary.

The once intranational debates over the Aral Sea problem (where extensive use of the water of the Syr Dar'ya and the Amu Dar'ya, fertilizers and defoliants to grow cotton has led to the shrinkage of the lake, loss of fish, destruction of agriculture, severe soil and wind erosion and terrible human problems) now requires international decision making (Micklin 1988; Glazovsky 1990). These are surely to be compounded if one solution the transfer of water from Siberia, became a reality (Figure 1.5a and b).

It is emphasized that a political viewpoint or comment is not being made here about the rights or wrongs of any of the examples. Instead it is intended to demonstrate that wherever decisions are made to manipulate huge natural systems there are sure to be geomorphological dimensions which may even be of global proportions. Many are World Bank or UNEP issues. Do we as international colleagues, with a very special and valuable knowledge, have a duty to draw attention to the scientific consequences? Perhaps this is not a role for an international association? I do believe, however, that we should work to equip those responsible with the standards and codes of practice to act responsibly on our behalf. This discussion therefore suggests another primary responsibility:

Responsibility 3. Trans-Frontier Issues and Good Neighbourliness

To work for good neighbourliness over trans-frontier issues by encouraging and equipping responsible agencies to properly recognize, research and implement the geomorphological dimension at both national and international levels.

To facilitate this it is also necessary to ensure that there is an adequate networking between human groups and cooperation between working groups for training and the transfer of information.

The European experience in the EC EPOCH and ENVIRONMENT Research programmes such as MEDALUS. The Blue Plan for the Mediterranean and the development of research centres such as the Centre for European Geomorphological Risks are examples of how we might influence events at the international scale.

THE GEOMORPHOLOGIST AS TECHNICIAN

To achieve all this the geomorphologist must be technically proficient across a wide field, including the correct use of remote sensing, air photographs and cartographic skills; the ability to observe, describe, measure, sample, collect and analyse spatial and temporal data; an understanding or ability to carry out the monitoring of the hydrologist, the investigative methods of the geophysicist and the laboratory methods and modelling of the soil chemist and soil mechanician. The addition of the geographer's awareness of the socio-economic domain and a willingness to use the documents of the planner add up to an impressive list. There is also the overall requirement to keep an open mind and a willingness to listen and learn from all the other sciences which might be involved (Jones 1980; Cooke and Doornkamp 1992).

This is a formidable and challenging task and it is clear that at all stages we are aiming at a professional product. To achieve this it is essential that the work is carried out to high professional standards, adheres to recognized codes of practice and is technically competent, A client demands the degree of skill which is expected of a competent practitioner in the profession (Sherrell 1975). Since the legal test of competence is "contemporary practice as judged by the standards of experts", it follows that applied geomorphologists must endeavour to maintain very high standards, indeed standards that most academics would do well to attain in their pure studies!

We all know, however, that we do not have a mechanism to deliver these standards nor do we have an agreed set of codes or a rigorous procedure for the establishment and upgrading of technical procedures. Although this may be considered by some as just the sort of bureaucracy that a geomorphologist and this Association wants to avoid, I feel that

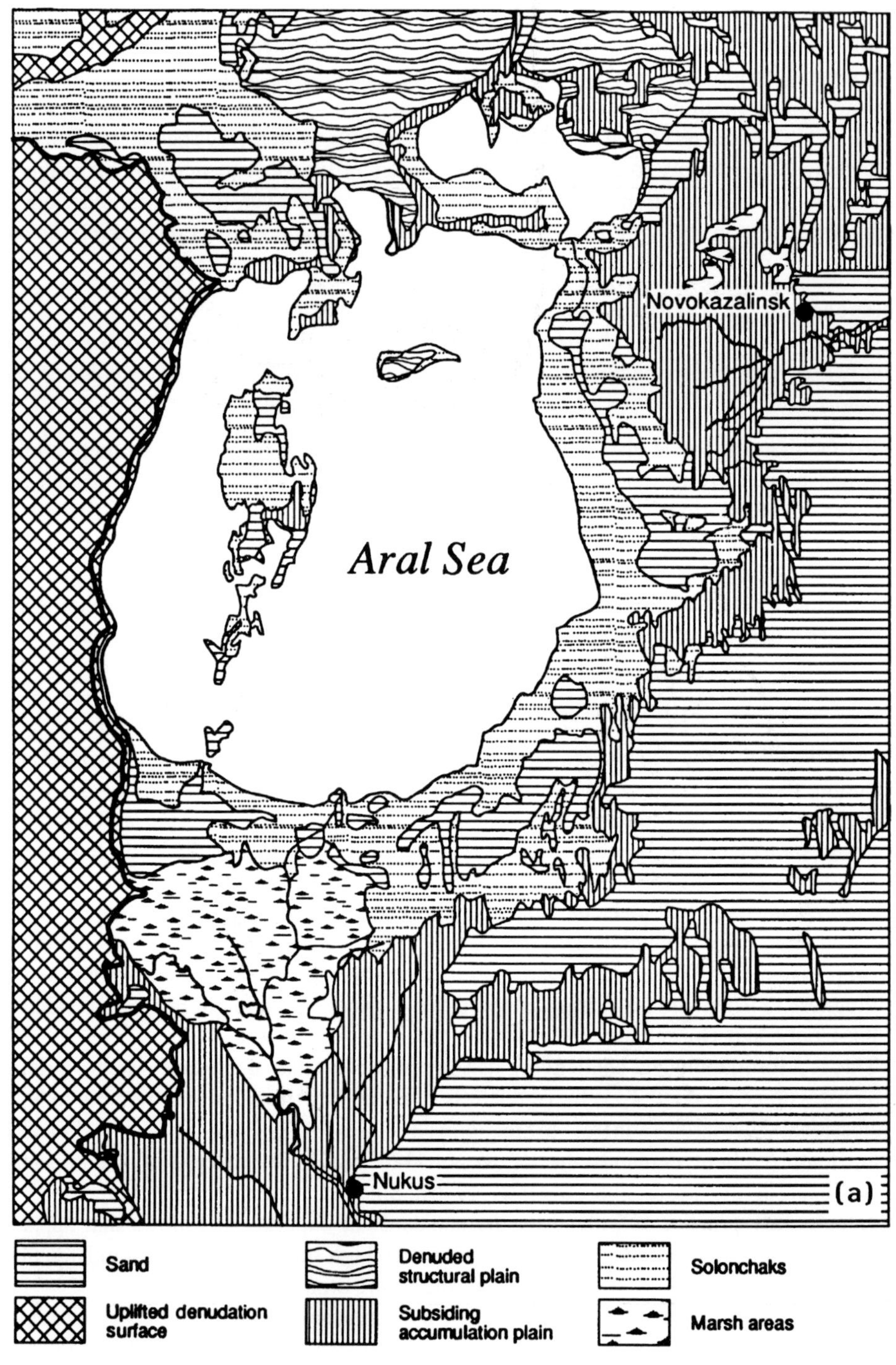

FIGURE 1.5 The Aral Sea. (a) Geomorphological setting. As supplied by Glavosky (1990)

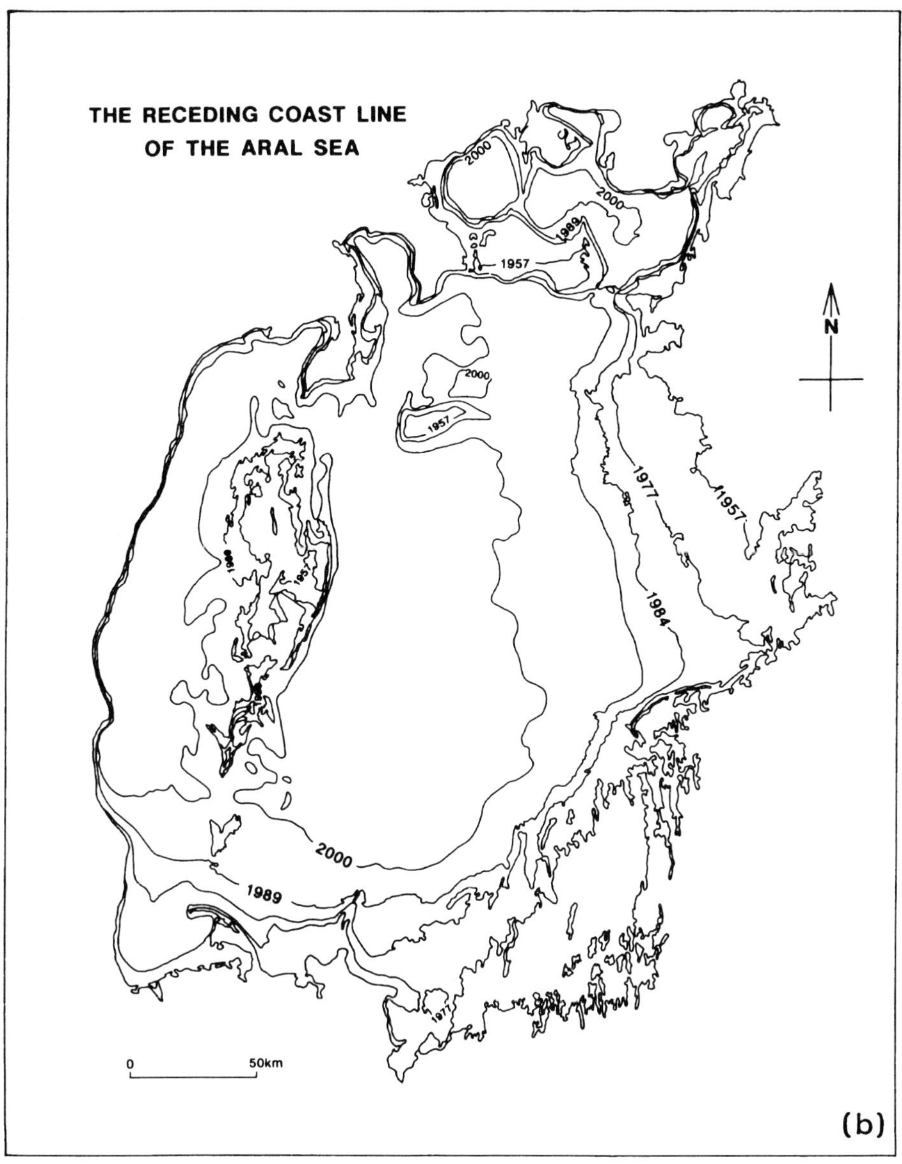

FIGURE 1.5 The Aral Sea. (b) sea-level change. As supplied by Glazovsky (1990)

the question should be explored. As a profession we are far behind medicine, law, engineering, surveying or geology, who have systems of watchdog committees, disciplinary councils, or Chartered Institutions. We are the only profession in the fields of engineering and science not to be represented by its own professional body even though all the evidence shows that these professions have been strengthened by the existence of a representative body and their public standing has been enhanced by professionalism. We do not have Standards Institutions such as the ASTM or the British Standards Institution. We do not have Codes of Practice such as the British Standards for Engineering Geology nor do we have a say in the standards by which we are given accredited or legal status.

ALTERNATIVES FOR PROFESSIONALISM

It would be wrong for me to pass lightly over these issues. At present we can only be represented or accredited through other disciplines. I will give three examples. In the UK this might be representation through the Royal Society for international affairs and accreditation through the Institution of Geologists (Geological Society of London), the Institution of Civil Engineers or the Institution of Mining and Metallurgy. In America it might be the American Institute of Professional Geologists. In Canada, based on a complicated range of provincial and territorial regulations, there is a major debate, with national registration efforts being coordinated by the Professional Registration Committee of the Canadian Geoscience Council. Geomorphologists are accredited under the title Professional Geoscientist.

This may be satisfactory for many who regard themselves as geologists but in other places geomorphologists may be geographers, pedologists or, as in Turkey, many are classed as engineers. There are very real problems in different parts of the world where the whole character of the discipline, its aims and its training patterns change as we move across frontiers. In Turkey we were asked for advice on whether geomorphology should be an independent department or whether it should be under science, engineering or geology. Here the jobs and status of the practitioner depends on the answer adopted and I am not sure that, at the time, we understood the implications of the questions or gave very good answers.

Again for many this is of no interest, but if we are serious about trying to help with the problems of the world we should recognize that we may not be accredited as we attempt to provide our advice in a different country from the one in which we received our qualification. Mobility is a prime reason for agreed codes. Consider, for example, the problems of establishing qualifications for interchangeable employment in the European Community in a minimum of 12 languages and university systems. All professional disciplines are currently negotiating this but there is no organization directly speaking for geomorphology.

Note the contentious issue that the course content of university and tertiary education might be influenced by these requirements. This must be a national-level decision because of the different priorities for the subject in each country but there are international requirements if geomorphologists are to work across frontiers. To put it bluntly, is a degree from one country acceptable in another? If not, can you accept an academic paper, a research result, or a consultancy report with professional confidence?

Within each national group we should perhaps be asking ourselves:

1. Do we want professional status?
2. If so, do we want to accredit, represent and train ourselves, to our own standards?
3. If not, do we wish to accept this from others? For example, should it be engineers or geologists or mining specialists who approve the geomorphology training and degrees?
4. Should there be international standards for the technical codes of practice?
5. What is the international role?
6. How can the IAG help?

In making these decisions the following overlapping and interlinked responsibilities might be borne in mind:

Responsibility 4. Standards and Quality Control
To establish, implement and improve the technical standards, codes of practice and quality of geomorphology.

Responsibility 5. Codes of Conduct
To establish a code of ethics and professional conduct and to define the relationship between the public and the professional geomorphologist.

Responsibility 6. Professional Accreditation
To encourage and facilitate the establishment of acceptable systems of training, education and professional experience to establish the criteria for the accreditation of competence and the international acceptance of individual national standards.

Responsibility 7. Representation and Influence
To represent geomorphological interests at government level, to sponsor international links, to influence and advise government, decision makers, legislative and technical committees and to fulfil a public relations role where there is a geomorphological interest or concern.

Responsibility 8. Dissemination of Information
To prepare, publish and distribute information of geomorphological and professional significance, including the Codes of Practice, Standards and Ethics.

Here we should draw attention to the fact that the IAG is already committed to the transfer of technical and academic information across frontiers. We have initiated the establishment of the Jan de Ploey Libraries as a permanent memorial to our late colleague, whose idea it was. The libraries are there to act as repositories of books and papers on geomorphology from all over the world and thereby to make available texts, in regional centres, in each continent. Achievements such as this are perhaps a primary function of an organization such as ours.

IS PROFESSIONALISM A BAD THING?

It is clear that geomorphology, whether we would prefer to be enchanted by long-term conversations with earth or not, is increasingly affected and necessarily committed to a pragmatic role. It is true that we are affected by convention. Research is funded by government, private University sources, grants from Charitable trusts, international agencies, the military, commerce and industry and some of these do have a solely practical purpose. Considerable funds are allocated by society to the professional institutions, the research stations, institutes of oceanography or hydrology, geological surveys etc., which are charged with a public duty to give advice on practical and environmental problems. In times of financial constraint these trends become intensified.

The question arises, "Is this a bad thing?"

To examine whether practical work poses a threat to the fundamental work of the subject it is necessary to examine whether applied geomorphology has inhibited the growth of theory, whether it has contributed to new theory and whether it adds substance to theory that has emerged in other ways. A powerful test of ideas is whether they work when put to the practical problem? The best form of hypothesis testing is to try to solve a practical problem in the natural landscape laboratory. Deficiencies which may cost lives are quickly revealed and there is a strong stimulus to discover more appropriate models.

We quickly discover that applied geomorphology has been responsible for the development of many new ideas. The best early examples are found in fluvial geomorphology and hydrology where much of our understanding of regime theory and hydraulics was developed by river and canal engineers (e.g. Lacey 1929–30; Lane 1955; Langbein 1962). Similarly, concepts of the passive behaviour of earth materials, loading, limit equilibrium and slope stability developed for practical reasons (e.g. Collin 1846; Terzaghi and Peck 1948). The classic works on soil erosion by Horton (1933, 1945), on sediment transport by Hjulstrom (1939) or on wind erosion by Chepil (1945), provide strong confirmation. Even the latest developments in slope stability, using hillslope hydrology, were developed out of contract research in Hong Kong and elsewhere (e.g. Anderson 1983). Similar examples can be cited from all aspects of the subject (see Brunsden et al. 1978; Cooke and Doornkamp 1992).

Applied work has also influenced the development of the great paradigms of our time. During the late 1960s and 1970s the cyclical and equilibrium models came under scrutiny as both pure and applied work showed that many systems were in disequilibrium and that unsteady behaviour in systems deserved wider attention. Today the dominant concepts of the subject stress the following keywords: thresholds, transient behaviour, episodicity, relaxation time, reaction, recovery, dominant and formative events, neocatastrophism, neotectonics. Throughout science attention is being paid to catastrophe, bifurcation, chaos, punctuated equilibria and extinctions. It is relevant to note that geomorphological applications are often concerned with disequilibrium, thresholds of change, recovery to a more stable state, disaster. Much of the stimulus and many of the examples for this theoretical development have derived from the need to find solutions to practical, unsteady and formative event situations. At the least we can argue that applied work is in sympathy with the aims of fundamental science.

There is no indication that it has inhibited the subject.

The huge funding that often accompanies applied work also means that the geomorphologist has powerful technology and logistic support to accomplish the project. This has had an abundant spin-off for the subject as a whole. For example, the widespread use of remote sensing to evaluate (using geomorphological methods) the structural framework of a potential hydro-carbon prospect is also valuable for studies of long-term landscape evolution. The employment of side-scan sonar and multi-beam bathymetry to map the ocean floor has had an innovative effect on our knowledge of submarine landsliding. There are innumerable examples from hazard mapping, the measurement of hazards on alluvial fans, the assessment of flow in wadis, sediment yield modelling, hardware simulation studies and beach forecasting, where applied work has advanced the subject.

There are two other areas where the pragmatic purpose has significant geomorphological benefits and I will now deal with these.

1. HUMANS AS GEOMORPHOLOGICAL AGENTS

The recognition that human beings can be potent agents of geomorphological change has been recognized by many authors (Goudie 1981). Generally these comments have drawn attention to the damage and the changes that we have made under the heading anthropogenetic geomorphology. The effect of forest clearance on soil erosion, sedimentation and flooding; the dissociation between humans and the soil; the indiscriminate use of resources; the effects of forestry, grazing, agriculture and industry including estimates of erosion and sedimentation are good examples.

They are, however, comments on the human impact on the landscape and always have an implication that our work is dangerous or wrong, rather than including examples of where we got it right. It seems important to spend considerable research effort in developing practical methods of landform design which are compatible with or improve upon the natural landscape. Examples include channel modification, slope design, beach feeding and quarry restoration. As Cooke (1992) has pointed out, the need is for sympathetic environmental geomorphology in which we ensure the environmental harmony of a project by including geomorphology as a "material consideration" in the planning process.

We should have confidence that we *do* understand landscape and that we can advise engineers on the forms and processes which will allow their designs to merge imperceptibly into the existing domain. It ought to be recognized that human beings have successfully created numerous beautiful and functional scenes, many of which we now fight to preserve.

2. INADVERTENT MODIFICATION OF PROCESS AND FORM

The second area of benefit we should recognize is the possibility of avoiding the inadvertent modification of geomorphological form and process. It has to be admitted that much of what we have got right has often been due to chance or due to the tolerance of the system. In many cases, however, because we do not thoroughly understand the system, quite unexpected or disastrous results may occur. Let me illustrate with an example from my home area.

The West Dorset Coast, in the UK, is famous for its beauty, its landslides and the Chesil Beach. The beach was produced at the end of the last glaciation by the rising sea as it drove onshore the Holocene beach. Once onshore it became reliant on the erosion of the cliffs for new gravel which moved alongshore to the east with the dominant wave direction. It also became a barrier to human use.

FIGURE 1.6 West Bay Harbour, Dorset. (a) 1886, (b) 1985

In 1866 an engineer called Sir John Coode recommended the construction of two solid harbour piers to allow fishing boats to cross the pebble bar at a town called West Bay (Figure 1.6a). All seemed well because pebbles could bypass the mouth in both directions.

Unfortunately, human action and landsliding in the west, formed episodically effective barriers to pebble movement. The supply was, in fact, pulsed and not continuous as is often assumed in longshore drift studies. It was also, in the long term, less than the amount required to maintain the net flux, across the harbour mouth, at the transport capacity of the wave climate. As time passed the western beach at West Bay diminished, the cliff eroded and remedial measures had to be used, again and again.(Figure 1.6b).

A classical picture emerged of beach material accumulating on one side of the barrier and erosion on the other! A textbook example you might think, and on that basis the design of a new harbour, beach management and remedial measures seem to have been based. There is, however, something wrong. The material is on the wrong, updrift, side of the barrier. The unexpected had happened.

The reason is, of course, obvious if we consider the system (Figure 1.7). The textbook case only works if the system is open and capable of meeting the required flux of sediment. In this case there were barriers to sediment supply. At West Bay, however, the net drift to the east had continued but was increasingly unsupplied. It was enough to maintain the Chesil Beach to the east of the barrier which gave the false impression of "no change" and of a westward drift because the downdrift loss was being made good. The "accumulation" on the downdrift side was not what it seemed. In 1987 and 1989 massive south westerly storms hit the coast. Now, however, there was no more shingle to move east and resupply the downdrift area. That, too, began to diminish. A sudden and dramatic fall in beach level was the result. It may be of more than passing interest to international geomorphologists that now even the world-class Site of Special Scientific Interest of Chesil Beach is of finite volume and under attack (Bray 1990; Brunsden 1992, Brunsden and Goudie 1993).

The lesson is clear, we must try to understand natural systems in the human context before design or management decisions are taken and there is an international responsibility for World Heritage Sites.

INTERDISCIPLINARY RESEARCH AND RELATIONS

One of the powerful driving forces for the development of applied studies has been the need for geomorphology to re-examine its position at the interface of the earth, environment, social and economic sciences. The need to be more relevant, either to society or to geography, one of its parent disciplines, has prompted many geomorphologists to work in areas where they can contribute to the interests of society, planners and environmentalists as well as engineering practitioners and others in the "real world".

It is obvious that geomorphological applications are in harmony with the public and political interest in environmental management problems. Geomorphologists are now required to have an understanding of the socio-economic and technical planning of their surveys so that they are fully compatible with the needs of the socio-economic disciplines and especially those of urban planners. As Jones (1983) has eloquently argued, there is a clear area, the human–environment arena, where the geosciences and the socio-economic political sciences overlap and there is a need for research in what he defined as:

Natural hazard impact — to change planning and management practices and influence policy formulation.
Environmental monitoring — to assess changes in natural environment systems and evaluate the need for new practices.
Resource assessment — to provide the basis for development.
Impact Assessment — to predict future changes, including environmental impact statements and impact avoidance assessments.
Ex post audits or ex post assessments — to review in hindsight the accuracy of prediction, design and the success of projects and policies.

The strengthening of links between the earth and social sciences is one possible future for the discipline.

A second element of interdisciplinary work is the continuing awareness that there is much to be gained from the other earth sciences — ecology, soil science, Quaternary geology, geology, geophysics, hydrology, sedimentology — and from engineering, soil and rock mechanics. There is still much to be gained by the incorporation of the methods and numerical rigour of these disciplines. I repeat the message here, even though it was first stated many years ago, because it is so important. For example, in the prestigious Rankine Lecture to the Institution of Civil Engineers under the title "Geology, geomorphology and geotechnics'' Henkel (1982) made two flattering but sincere statements (pp. 177 and 193):

> The geological environment is complex with so many facets that control its behaviour that the only way to achieve a full understanding is for there to be an interdisciplinary approach in which engineers and geologists work much more closely together. The meeting ground can, I believe, be found by both the professions concentrating on understanding the geomorphology of construction sites

and,

> Over the years I have been involved in a wide variety of construction projects and those which have proved to be the most demanding and stimulating and have contributed most to my education have always been associated with the need to bring together geology, geomorphology and engineering ... My message is that, in order to define the fundamental assumptions, there is no substitute for painstaking study in the field of geology and geomorphology. We still need something of the Victorian virtue known as "eye for the ground".

FIGURE 1.7 (*opposite*) The nature of longshore drift when affected by human interference. (a) The open system. (b) Insertion of a partial barrier, e.g. a harbour mouth. (c) Insertion of a total barrier on the updrift (supply) side. (d) Depletion takes place on the updrift (wrong!) side of the harbour mouth but may be enough to maintain the beach on the downdrift side. (e) Supply beach store empties, there is no transport across harbour mouth and depletion now begins on the downdrift (correct) side of the partial barrier. Note that the model assumes that the cliffs behind the supply beach do not supply suitable material to make a beach which is realistic for the clay cliffs of the example discussed. The heavy arrow depicts the dominant drift direction

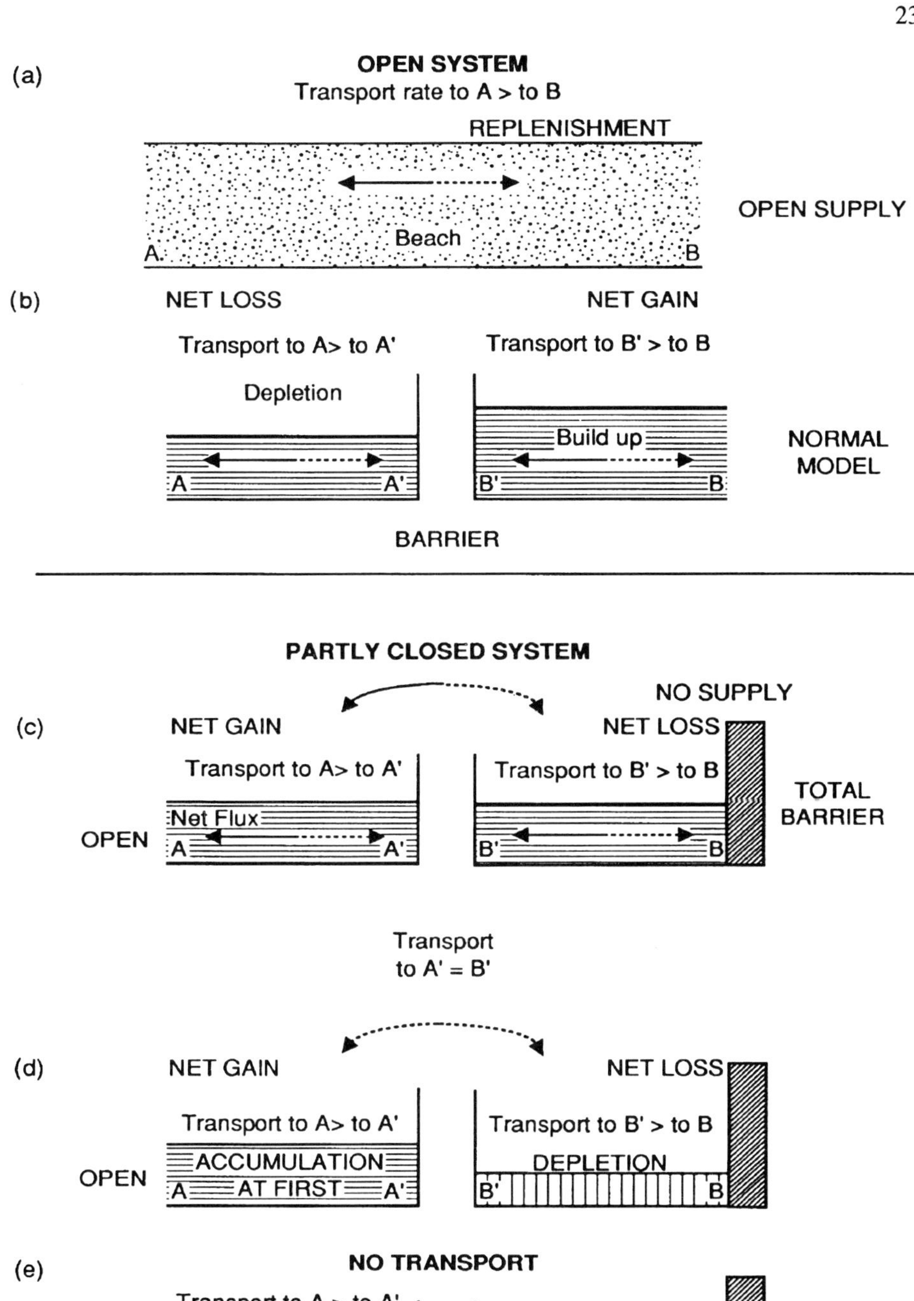

FIGURE 1.7 *For caption see opposite*

It cannot have escaped attention that the tender documents for international environment and development projects issued by such agencies as the UNEP, UNDP, World Bank, Asian Bank or national agencies now include clauses requiring an understanding of the geomorphology of a development site. The documents usually require that a knowledge of the earth science processes, materials and landforms be utilized in planning, land zoning, design of structures, building ordinances, codes of practice and remedial works. Geomorphological knowledge is now being utilized as an invaluable input to the success of the planning, development and construction process with a consequent improvement to the success of the project, to the level of aid we offer to the disadvantaged, to the steady avoidance and mitigation of hazards and to the wise use of the earth as the home of humans.

It is a salutary thought to realize that the agencies are asking for the inclusion of sound geomorphological science but we do not have in place the Codes of Conduct or Standards that seem to necessary. It is also important to realize that those who make decisions which affect the environment have to do so whether they receive geomorphological help or not. It is a valid view to suggest that the geomorphological input should be made by those best able to provide it — the geomorphologists — and not by other scientists, however well meaning or competent in their own fields they may be.

Cooke (1992) has pointed out another dimension of this theme. He showed that the three current imperatives of environmental geomorphology are:

— *the landscape imperative* or the interrogation and environmental audit of a landscape prior to development.
— *the institutional imperative* or the way in which we give environmental advice within institutional contexts.
— *the historical imperative* or the lessons which can be learnt from the historical record of (global) environmental change for the wise management of earth.

As he said, "Too often scientists fail to understand the complex and restricting institutional contexts of their research and managers fail to appreciate the complexity and limitations of environmental data."

The landscape imperative advocates a practical purpose for at least some of our conversations with earth and the institutional imperative covers the responsibilities listed as numbers 3–8 in this paper. All imperatives also cover the need for work at the interface of the human–environment milieu and the historical imperative directs our attention at the need to contribute to issues of profound importance to the well-being of the planet.

It is vital that we begin as an Association, to contribute to such global initiatives as the Global Change Programme (ICSU), the Human Dimensions of Global Change Programme (ICSU) and the International Decade for Natural Disaster Reduction (UNEP), and it is important that the IAG has agreed to a Task Force under the Chairmanship of Professor J.B. Thornes to guide us toward this. In this way I believe we will adapt to the needs of society and grow in status and responsibility. My last "Responsibilities" for the Association therefore are:

Responsibility 9. Institutional Cooperation and understanding
To encourage interdisciplinary training and work at the interface of the earth,

environment, socio-economic. political and planning disciplines in order to ensure understanding and cooperation within institutional frameworks.

Responsibility 10. The Wise management of Earth
To contribute a sound geomorphological dimension to both global and local programmes for the wise management of the process systems and resources of earth and the avoidance and mitigation of its hazards.

CONCLUSION

> It is part of man to speak for man and not to assert values that are dubiously assumed to inhere in his natural environment. And, in speaking for man he will find that the environment speaks for itself. (Curnow 1941)

In 1989, in Frankfurt, I offered, for discussion, ten propositions as a framework for the theoretical development of the subject (Brunsden 1990). I am very flattered that you have taken them seriously and been generous enough to debate them at this meeting. Today I place before you in the same spirit a possible agenda for our discipline as a profession, in the form of ten responsibilities for an emerging international association.

Responsibilities

1. *Equal Opportunity*. To work for the equal opportunity of all people to cross the frontiers of knowledge.
2. *Scientific Freedom*. To support the ideal of the scientific freedom to study without the pressures of convention, politics or oppression.
3. *Trans-Frontier Issues and Good Neighbourliness*. To work for good neighbourliness over trans-frontier issues by encouraging and equipping responsible agencies to properly recognize, research and implement the geomorphological dimension at both national and international levels.
4. *Standards and Quality Control*. To establish, implement and improve the technical standards, codes of practice and quality of geomorphology.
5. *Codes of Conduct*. To establish a code of ethics and professional conduct and to define the relationship between the public and the professional geomorphologist.
6. *Professional Accreditation*. To encourage and facilitate the establishment of acceptable systems of training, education and professional experience, to establish the criteria for the accreditation of competence and the international acceptance of individual national standards.
7. *Representation and Influence*. To represent geomorphological interests at government level, to sponser international links, to influence and advise government, decision makers, legislative and technical committees, and to fulfil a public relations role where there is a geomorphological interest or concern.
8. *Dissemination of Information*. To prepare, publish and disseminate information of geomorphological and professional significance including the Codes of Practice, Standards and Ethics.
9. *Institutional Cooperation and Understanding*. To encourage interdisciplinary training

and work at the interface of the earth, environment, socio-economic, political and planning disciplines in order to ensure cooperation and understanding within institutional frameworks.

10. *The Wise Management of Earth.* To contribute a sound geomorphological dimension to both global and local programmes for the wise management of the process systems and resources of earth and the avoidance and mitigation of its hazards.

I began this address with both a happy and a tragic event — a sad and distressing view of the world. I would not like that to be the memory that you take away for the next four years until we meet again. This has been an excellent conference. Throughout we have been looked after by a team who have demonstrated their concern for *our* well-being. They have also treated us to spectacular feasts of landscape scenery in the field, in picture competitions, in lectures and exhibitions. Let me end in the same vein as I thank them, Derek Ford, Brian McCann, Susan Vajocki and all the team on your behalf.

In doing so the opportunity still arises to emphasize that, even when attempting the purest of artistic or scientific endeavours, there will be a sense of responsibility to which I believe we should listen and respond. For example, my decision to become a geomorphologist stems from the kindness of an old lady, a Miss Whitacker, of Torquay, Devon, UK, who gave me discarded copies of the *National Geographic Magazine*. Scenes of Brice Canyon, the Arches National Park, Formosa (Taiwan), Niagara, coasts and mountains filled my young mind with the wonder of a beautiful world.

To this day it is this geomorphological "sense of place'' which brings most inner peace. I believe this to be true of all of us. We all have a special place. I have two, the Southern Alps of New Zealand and the Dorset Coast, but even in these places while enjoying the landscape imperative I cannot escape the human perspective. The former place has been described by a New Zealand authoress, Jill Tremain (1973): "What a privilage" she said, "to know the profound stillness and peace of the land".

Surely there could be nothing simpler or purer than to experience "The way in which nature and landscape influence the human imagination." It is thought-provoking therefore to find that the author of these words, E.M. Relph (1976) found it necessary to add, "A knowledge of landscape is a profound human experience that can only be fully realised by a wholehearted commitment to and caring for our world".

If we think of these landscapes only as geomorphological scientists, our experience can only be academic knowledge in which we seek understanding as a duty and where the "commitment'' is made because the earth is intrinsically interesting. I am sure that all members of our profession feel some sympathy with this view. We can never forget, however, that the earth is our home and the words "caring for'' ensure that the practical riposte is never far away.

This type of dichotomy has never been better expressed than by Samuel Butler (1872) in the Utopian novel, *Erewhon*:

> Suddenly as my eyes got on a level with the top, so that I could see over, I was struck almost breathless by the wonderful mountain that burst on my sight. The effect was startling ...

and then

> I am forgetting myself into admiring a mountain which is no use for sheep. This is WRONG. A mountain here is only beautiful if it has good grass on it. Scenery is not scenery — it is

"country", subaudita voce "sheep". If it is good for sheep, it is beautiful, magnificent, and all the rest of it, if not, it is not worth looking at. I am cultivating this tone of mind with considerable success, but you must pardon me for an occasional outbreak of the old Adam.

Could it be that when we are looking at the most "pure" aspects of our subject, the interaction between the human imagination the cultural and social experience, the aims of the individual and community, and the nature of the physical environment does not allow us, even then, to divorce the pure experience from the pragmatic requirement?

Thus, in response to the debate initiated by Baker and Twidale (1991) I look forward to the day when we can ensure that all people will have the freedom to hold "conversations with the earth" and that geomorphologists will truly be without frontiers. Until then perhaps we should pay more respect to our responsibilities!

ACKNOWLEDGEMENTS

I wish to acknowledge the help, companionship, generosity and friendship of all those geomorphologists who have been so kind as to help me in MY *conversations with earth* and who have supported our efforts to develop the International Association of Geomorphologists.

There are, of course, special people whose support has been vital. I do not name them because there are too many and I would be sure to forget someone. I therefore only record my thanks to my wife Elizabeth who has never missed an IAG meeting and who has been totally supportive.

Personal thanks are due to Robert Allison, Dietrich Barsch, Ron Cooke, John Doornkamp, Andrew Goudie, David Jones, Olaf Slaymaker, Jesse Walker, Roma Beaumont, Gordon Reynall and Peter Howard who have directly contributed to this paper and improved it by their comments.

King's College has been very generous with time and financial backing without which many IAG journeys would have been impossible.

Above all I thank John Thornes for his lifelong support, intellectual stimulation and interest in my activities.

REFERENCES

Anderson, M.G. 1983. *The Prediction of Soil Suction for Slopes in Hong Kong*. Final Tech. Rept. WD CE3/81. Geotechnical Control Office, Hong Kong, 242 pp.

Baker, V.R. and Twidale, C.R. 1991 The reenchantment of geomorphology. *Geomorphology*, **4**, 73–100.

Bray, M. 1990. Landslides and coastal zone sediment transport. In Allison, R.J. (ed.), *Landslides of the Dorset Coast*, Brit. Geom. Res. Group, Field Guide, pp. 107–117.

Bronowski, J. 1979. *The Common Sense of Science*. Harvard University Press, Cambridge, Mass., 154 pp.

Brunsden, D. 1990. Tablets of stone: towards the ten commandments of geomorphology. *Z. Geomorph. Suppl. Bd*, **79**, 1–37.

Brunsden, D. 1992. Coastal and landslide problems in West Dorset. *Papers and Proc., SCOPAC, Coastal Instability and Development Planning, 24 Oct. 1991, Portsmouth*, pp. 19–44.

Brunsden, D. and Goudie, A.S. 1993. *Classic Landforms of the Dorset Coast*, 2nd edition.

Brunsden, D., Doornkamp, J.C., Fookes, P.G., Jones, D.K.C. and Kelly, J.M.H. 1975a. Geomorphological mapping techniques in highway engineering. *Journ. Inst. Highway Engineers*, **22**(12), 35–41.

Brunsden, D., Doornkamp, J.C., Fookes, P.G., Jones, D.K.C. and Kelly, J.M.H. 1975b. Large scale geomorphological mapping in highway engineering. *Quart. Journ. Engin. Geol.*, **8**, 227–253.

Brunsden, D., Doornkamp, J.C. and Jones, D.K.C. 1978. Applied geomorphology: a British view. In Embleton, C., Brunsden, D. and Jones, D.K.C. (eds), *Geomorphology: Present Problems and Future Prospects*. Oxford University Press, pp. 251–262.

Brunsden, D., Doornkamp, J.C. and Jones, D.K.C. 1979. The Bahrain Surface Materials Resources Survey and its application to planning. *Geogr. Journ.*, **145**, 1–35.

Butler, S. 1872. *Erewhon or Over the Range* (in press). London. (Many reprints, including a Penguin edition.)

Chepil, W.S. 1945. Dynamics of wind erosion: II. Initiation of soil movement. *Soil Sci.*, **60**, 397–411.

Chorley, R.J. 1978. Bases for theory in geomorphology. In Embleton, C., Brunsden, D. and Jones, D.K.C. (eds), *Geomorphology: Present Problems and Future Prospects*, Oxford University Press, 1–13.

Cloos, H. 1953. *Conversation with the Earth.* Knopf, New York, 413 pp.

Collin, A. 1846. *Glissements spontanes des terrains argileux.* Carilan-Goeury et Dulmont, Paris. English translation: Schreiver, University of Toronto Press, 1956.)

Cooke, R.U.C. 1987. The use of geomorphology. In Gardner, V.(ed.), *International Geomorphology, Part 1.* Wiley, Chichester, pp. 63–82.

Cooke, R.U.C. 1992. Common ground, shared inheritance: research imperatives for environmental geography. *Trans. Inst. Brit. Geogr.*, **5**(17), 131–151.

Cooke, R.U.C. and Doornkamp, J.C. 1992. *Geomorphology in Environmental Management: A New Introduction*, 2nd edition. Oxford University Press, 410 pp.

Cooke, R.U.C., Brunsden, D., Doornkamp, J.C. and Jones, D.K.C. 1982. *Urban Geomorphology in Drylands.* Oxford University Press, 324 pp.

Curnow, A. 1941. *The Unhistoric Story, Recent Poems.* Caxton Press. (Quoted, p.13, in McNaughton, T. *Countless Signs: the New Zealand Landscape in Literature.* Reed Methuen, p.389.)

Doornkamp, J.C., Brunsden, D. and Jones, D.K.C. 1980. *Geology, Geomorphology and Pedology of Bahrain.* GeoBooks, Norwich, 443 pp.

Glazovsky, N.F. 1990. The Aral crisis: the source, the current situation, and the ways to solving it. *International Symposium Proc., The Aral Crisis: Oorigins and Solutions, Oct. 2–5 1990, Nukus, Karakalpak Autonomous Sov. Soc. Republic (Uzbek SSR)*, Academy of Sciences Moscow, USSR, 55 pp.

Goudie, A.S. 1981. *The Human Impact: Man's Role in Environmental Change.* Basil Blackwell, Oxford, 316 pp.

Goudie, A.S., Brunsden, D., Collins, D.N., Derbyshire, E., Ferguson, R.I., Hashmet, Z., Jones, D.K.C., Perrot, F.A., Said, M., Waters, R.S., and Whalley, W.B. 1984. The geomorphology of the Hunza Valley, Karakoram Mountains, Pakistan. In Miller, K.J. (ed.), *The International Karakoram Project Volume 2.* Cambridge University Press, Cambridge, pp. 359–410.

Henkel, D.J. 1982. Geology, geomorphology and geotechnics. *Geotechnique*, **32**(3), 175–194.

Hjulstrom, F. 1939. Transportation of detritus by moving water. In Trask, P.D. (ed), *Recent Marine Sediments.* American Association Petroluem Geologists, pp. 25–31.

Horton, R.E. 1933. The role of infiltration in the hydrological cycle. *Trans. Amer. Geophys. Union*, **14**, 446–460.

Horton, R.E. 1945. Erosional development of streams and their drainage basins: Hydrophysical approach to quantitative morphology. *Bull. Geol. Soc. Amer.*, **56**, 275–370.

Jennings, J.N. 1973. "Any milleniums today, Lady,'' the geomorphological bandwagon parade. *Aust. Geog. Studies*, **11**, 115–133.

Jones, D.K.C. 1980. British applied geomorphology: an appraisal. *Zeits. fur Geomorph., Suppl.*, **36**, 48–73.

Jones, D.K.C. 1983. Environments of concern. *Trans. Inst. Brit. Geogr., New Series*, **8**, 429–457.

Jones, D.K.C., Brunsden, D. and Goudie, A.S. 1983. A preliminary geomorphological assessment of part of the Karakoram Highway. *Quart. Journ. Eng. Geol.*, **16**, 331–355.

Lacey, J. 1929–30. Stable channels in alluvium. *Proc. Insti. Civil Eng.*, **229**, 259–292.

Lane, E.W. 1955. The importance of fluvial morphology in hydraulic engineering, *Am. Soc. Civil Eng. Proc.*, **81**, 1–17.

Langbein, W.B. 1962. Hydraulics of river channels as related to navigability. *U.S. Geol. Surv. Water Supply Paper*, **1539-W**.

Micklin, P.P. 1988. Desiccation of the Aral Sea: a water management disaster in the Soviet Union,

Science, **241**, 1170–1176.
Relph, E.M. 1976. *Place and Placelessness*. Pion Press, London.
Reynolds, M.R. and King, P.S.D. 1988. *The Expert Witness and his evidence*. B.S.P. Professional Books, Oxford, 137 pp.
Sherrell, F.F. 1975. Professional liability and professional indemnity for consultants. *British Geol.*, **1**, 35–6.
Terzaghi, K. and Peck, R.B. 1948. *Soil Mechanics in Engineering Practice*. Wiley, New York. (2nd edition, 1967).
Thornes, J.B. 1978. The character and problems of theory in contemporary geomorphology. In Embleton, C. *Geomorphology: Present Problems and Future Prospects*, Oxford University Press, pp. 14–24.
Tremain, J. 1973. Southern Alps winter ski traverse. *The Canterbury Mountaineer*, **41**, 43.
Walker, H.J. 1978. Research in coastal geomorphology: basic and applied. In Embleton, C. *Geomorphology: Present Problems and Future Prospects*, Oxford University Press, pp. 203–223.

2

The Great Debate: Denys Brunsden's Tablets of Stone

Contributions by Olav Slaymaker, Hanna Bremer, C.R. Twidale, Eiju Yatsu, Victor Baker, Jean L.F. Tricart and Denys Brunsden

INTRODUCTION AND CONTEXT

OLAV SLAYMAKER

In recent years, several authors have set forward basic propositions for the conceptual and theoretical framework of our subject. For example, in 1985 Stanley Schumm discussed seven reasons for geologic uncertainty and in 1987 Adrian Scheidegger published his fundamental principles of landscape evolution. At the Second International Geomorphology Conference held in Frankfurt in 1989, Denys Brunsden prepared ten propositions, which he called "Tablets of Stone" and invited all geomorphologists to offer their own alternatives (Brunsden 1990). The Programme Committee of the Third International Conference of Geomorphologists thought that it would be valuable to respond to Brunsden's challenge by inviting five distinguished geomorphologists to prepare comments to serve as a basis for a plenary debate at the Hamilton Conference. All five colleagues responded with written papers — Vic Baker (USA), Hanna Bremer (Germany), Jean Tricart (France), C.R. Twidale (Australia) and Eiju Yatsu (Japan); unfortunately, Jean Tricart was unable to attend the conference. This contribution introduces Brunsden's propositions, gives brief summaries of the debaters' positions and adds a comment on the plenary discussion that ensued.

Brunsden (1990) defined ten propositions as a theoretical framework for geomorphology. They are:

1. The style and location of landform change is determined by the type, location and rate of tectonic movement.
2. Landforms are shaped by tectonic and denudation processes proceeding concurrently.
3. The lower boundary for landform development is set by varying sea levels.

Geomorphology Sans Frontières. Edited by S. Brian McCann and Derek C. Ford.

4. For any given set of environmental conditions there will be a tendency to produce a set of characteristic landforms.
5. Landforms are continually subject to perturbations. These impulses are episodic and complex.
6. Within each tectono-climatic regime, landforms are produced by specific process events called formative events.
7. New landforms are produced when an event is reached on the frequency–magnitude scale of a given tectono-climatic regime and a geocatastrophe occurs.
8. When a perturbation exceeds the resistance of the system, the system will react and relax toward a new stable state or characteristic form.
9. There is wide spatial variation in landform sensitivity to change. Hence landscape stability is diverse and complex.
10. The ability of a landscape to resist impulses of change tends to increase with time.

These ten propositions were summarized by Brunsden in the five "cardinal propositions" of

1. complex cause (3, 5, 6 and 7)
2. complex response (8)
3. sensitivity to change (9)
4. tectono-climatic interaction (1, 2)
5. temporal resistance to change (4, 10)

It is also possible to summarize the ten propositions as stating the role of

1. structure (1, 2 and 3)
2. process (5, 6, 7, 8 and 9)
3. stage (4 and 10)

Bremer reviewed the first three Brunsden propositions and demonstrated, with examples, the ways in which these principles are equally fundamental to process geomorphology and climatic geomorphology. She indicated that Brunsden's paper represented a valuable basis for linking the various national schools of geomorphology. Twidale reviewed Propositions 4, 6 and 10. While he did not take direct issue with much of the substance of these propositions, he questioned the wisdom of trying to develop general laws about landscape development on the empirical grounds that general laws always miss contingent realities. He also noted that the ubiquitous role of water was given inadequate attention by Brunsden. His most distinctive contribution to the debate concerned the implications of very old paleoforms for general theory. Yatsu reviewed Propositions 5, 7, 8 and 9. He also thought that the substance of individual propositions was largely acceptable, though he objected to the term "geocatastrophe" in Proposition 7. His main concern was with the aims, objectives and framework of the whole paper which he characterized as being "too Davisian". Baker took issue with the assumption that cardinal propositions are (a) possible and (b) important. His view is that "the Golden Rule is that there are no golden rules" (Baker, this volume, p. 61), a position which is supported "from the logic of science, from pragmatic philosophy and from common sense". He stated that the unending quest for truth is more important than the statement of a neo-Davisian theoretical framework. Tricart's discussion takes a broad systems approach and looks at internal geological influences, external forces and productive human activity as the three axes of geomorphology. He concluded with his well-known views on the characteristic time scales of components of terrestrial natural systems. His most serious concern with Brunsden (1990) is that a listing of commandments tends to overlook the interrelationships between components of geomorphic systems.

There were therefore three kinds of critique of Brunsden (1990):

(i) the inadequacy of individual propositions (all except Baker included some refinements);
(ii) the inappropriateness of the Brunsden framework (Yatsu and Tricart); and
(iii) the inappropriateness of attempting to create a framework (Baker and Twidale).

In the discussion from the floor which followed the four introductory statements by Bremer, Twidale, Yatsu and Baker, there were 14 interventions which dealt with topics as disparate as the nature of truth, the importance of the Pleistocene and whether nature recognizes the usefulness of scientists' classificatory schemes. In reflecting on the event it seems clear that Tricart's absence was serious, as he was the only one to offer a viable alternative framework to that of Brunsden. Many of the intervenors from the floor asked about the role of human activity, the landscape–soil–management interface and applied geomorphology. In this context it is interesting to note that Brunsden's Presidential Address at Hamilton was a brilliantly focused statement on the societal obligation of geomorphologists.

As I came away from the crowded lecture theatre I overheard many heated conversations. One said "I hope they never try making a silk purse out of a sow's ear again" and another said "Just what geomorphology needs — a re-examination of our fundamental premises". The great debate will continue whether we wish to participate or not.

REFERENCES

Brunsden, D. 1990. Tablets of stone: toward the Ten Commandments of Geomorphology. *Zeitschrift fur Geomorphologie, Supplement Band*, **79**, 1–37.

Scheidegger, A. 1987. The fundamental principles of landscape evolution. *Catena Supplement*, **10**, 199–210.

Schumm, S.A. 1985. Explanation and extrapolation in geomorphology: seven reasons for geologic uncertainty. *Transactions of the Japanese Geomorphological Union*, **6**(1), 1–18.

CLIMATIC, CLIMATOGENETIC AND TECTONIC GEOMORPHOLOGY

HANNA BREMER

The first three proposals of Brunsden (1990) are concerned with the influence of tectonics, base-level and climate on landforms.

In the first proposal a linkage of tectonic and climatic geomorphology is considered. Tectono-climatic regions are outlined by overlaying a morphoclimatic and a tectonic map. The morphoclimatic map gives the zones of recent relief-forming mechanisms, while the tectonic map outlines orogenic (according to age) and anorogenic regions (subdivided into plateau basalt, platform and shield). Thus the second map says something about the structure of the rocks, and refers only indirectly to slow uplift and subsidence (in the following referred to as uplift). A similar map is constructed for 18 000 years BP. Comparing those two maps Brunsden defines regions of stability and areas which have changed significantly but still have a strong relict input.

Perhaps one would like to have different boundaries, but this is not a point for a general critique. On the whole this proposal is a valuable framework for detailed investigations. The map of climato-tectonic regions is derived by deduction. It gives an overall idea of the relief-forming mechanisms. This is most important for evaluating the significance and interrelation of recent processes. These may be measured or observed.

Here is a way to bridge the gap between process geomorphology and climatic geomorphology, as the latter usually works on a longer time scale. This would be the basis then for meso- and mega-geomorphology which necessarily incorporates historical elements. The antagonism between process geomorphology and historical geomorphology is vanishing anyway as is shown by recent papers, e.g. Kennedy (1992) and Brunsden (1993). If you are measuring processes there is a necessity to provide a broader framework as Barsch (1990) he said already at the Frankfurt conference. For instance, you may fit detailed data into the framework proposed by Brunsden.

There are signs that preconceived ideas, systems and models, some taken from other sciences, are becoming less important. Dynamic equilibrium, steady state and steady evolution may well fade from the discussion as they are assumptions which are almost impossible to prove in nature. They may be applied only for those relief elements where the recent origin has been verified. For instance in my dissertation I tried to show that the Weser River is in equilibrium since under natural conditions no incision or accumulation which would change the profile has happened for 6000 years (Bremer 1959). However, landforms in most areas are predominantly relict forms which do not fit into concepts of equilibrium. These relict forms are not postulated but may be deduced from independent criteria like fossil soils.

Looking at Proposition 1 from the perspective of climatic geomorphology it may be a framework for further research. So far climatic geomorphology has been concerned with exogene processes. The interrelation of processes, the relief-forming mechanisms, are the aim of climatic geomorphology, but not the linkage of single processes with climatic data. Perhaps the alternatively used term "morphodynamic system"(Budel 1982, p.26) illustrates this even better.

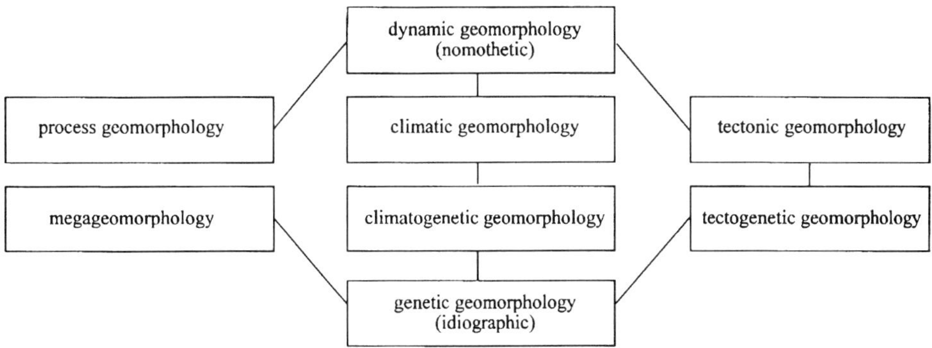

FIGURE 2.1 Different approaches to geomorphology. So far the influence of tectonics upon processes has been little evaluated. Usually landforms in their tectonic history are considered in what is called "tectonic geomorphology"

As there is sometimes a confusion in the English literature, short definitions may be given (cf. Figure 2.1). Climatic geomorphology is the investigation of recent processes in a morphoclimatic zone and their interrelations. Process geomorphology by contrast observes single processes and their climatic and hydrologic controls. It is easy to say there are "remarkably similar forms the world over"(Baker and Twidale 1991, p.81), but this does not account for their frequency or importance nor for their possible paleocharacter.

The ratio between denudation and fluvial erosion — i.e. incision or accumulation, not transportation — is fundamental in climatic geomorphology. Direct observations of processes are cross-checked by landform analysis. Processes are extrapolated and it is figured out whether this explains the landform in question. Sequences of similar landforms allow for interpolation and the derivation of causal processes. Increasingly, climatic geomorphology and process geomorphology are combined, either in measuring processes in climatic geomorphology (e.g. in Barsch 1990) or in trying to balance different measured processes with accumulations, as was demonstrated during the pre-IAG conference excursion in Canadian alpine environments in 1993 by Slaymaker and Luckman and their colleagues (Slaymaker 1993).

Process geomorphology and climatic geomorphology may be subsumed under the heading "dynamic geomorphology". These methods aim for nomothetic results and strive to discover rules about geomorphological processes. Exogene forces are assumed to be the major causative factors. We know little about the influence of endogene forces except at places where they are very rapid. There is little known about the influence of slow tectonic movements on geomorphological processes. Relative mountain heights and denudation are the basis for models cited in Brunsden (1990). Uplift is not a continuous movement, even more the reaction by geomorphological processes is discontinuous not only in the scale of years and centuries and millennia, but especially in space. Examples are the interruptions of erosion at rims (Budel 1982, fig. 10), and the storage of deposits in glaciated areas (Slaymaker 1992).

Climatic geomorphology brought about climatogenetic geomorphology, as in almost all regions of the world there are relict landforms of former climates more or less widely distributed. Climatogenetic geomorphology is idiographic. As a genetic science it does contain nomothetic aspects. In a defined region, it examines the evolution of landforms. The interfingering of relief elements gives the first clues for the classification of single forms according to relative age. Relief generations are deduced independently. Climatic causes are inferred by zonal comparison at a later stage. Rock resistance is evaluated and if possible abstracted. Tectonic movements are taken into account if they are known from geological indicators like dated sediments. As far as possible guesswork about tectonic movements is avoided. Climatogenetic geomorphology works on about the same scales as the morphostructures investigated by Russian colleagues. This is definitely a smaller scale than the morphoclimatic zones of Budel and it is also smaller than the "tectonic domains"of Brunsden (1990).

Climatic and climatogenetic geomorphology did not account very much for tectonic movements unless they were obvious. High mountains were excluded from Budel's map of zones of climatic geomorphology. Probably the assumption was that the tectonic influence outside of high mountains is of minor importance for recent processes. It may be time now to look for the differences of relief-forming mechanisms under different tectonic set-ups but under the same morphoclimatic conditions. One of the major methods of climatic geomorphology is the spatial comparison. Brunsden's maps outline those regions where other factors are equal but tectonics differ. Thus these maps may be a framework for field investigations.

There are problems of different climatic and tectonic history. The evolution of landforms over the last 60 or even 100 million years is only partly normative, partly ideographic. Looking at the map of tectonic–climatic regions (Brunsden 1990, fig. 1a–c) one should realize that there are well-defined time sections or climatic states delineated and not relief generations of climatogenetic geomorphology. A comparison may illustrate this. Though Central Australia and the central part of the Sahara might have had a similar climatic history, the landforms are very different. In Central Australia, due to very low relief, paleosols and paleolandforms from a (Tertiary) humid wet climate are widely preserved. Not so in the Sahara, where there are only few relicts from pre-Quaternary age. These differences also influence recent processes. It is not only climate and tectonics which control recent processes but inherited landforms, too. We need more well-investigated examples of relief generations which developed under similar climatic conditions to evaluate the influence of tectonics on processes.

Proposition 2 (Brunsden 1990) is equally important — what is the velocity of uplift and how fast is the reaction of the processes and how fast are the landforms shaped? This is obviously a question of scale as Brunsden (1993) just pointed out. His Proposition 2, that tectonic and denudation processes proceed concurrently, is certainly true, but at what rate? We cannot assume that landforms directly reflect the ratio of endogene to exogene processes. This was the basic concept of Walter Penck but like the concept of Davis it is not verified in nature. Both concepts rely ultimately on gravity. The higher a point the higher the potential energy, the steeper the gradient, the larger the erosional power. This neglects friction and inertia which are quite different in spatial and temporal dimensions mainly due to weathering. In the tropics, for instance, steep rock walls may exist for millions of years (e.g. Ayers Rock).

If uplift and denudation rates are compared in a direct more or less steady relation, only gravity is taken into account. Climatically induced processes in all their variety, as well as petrovariance, both largely controlled by friction, are neglected. Furthermore, there is discontinuity of uplift and there is a very different relaxation time in landform shaping. Thresholds and inertia have to be overcome. All this is due to friction (Bremer 1989). Difference in erosion rates may occur with the size of the discharge area due to storage (Godard 1990). This is most important too in mountain basins (Slaymaker 1992).

What do we know about the influence of uplift so far? For geomorphology the total amount and the speed of uplift have to be investigated as both influence the processes. The style of uplift (wide folds = cymatogeny, or faulting) is important for spatial changes. The total amount of uplift above sea level may be deducted from the highest points in a landscape. If there are relicts of planation surfaces in the crest areas we can conclude that this is a remnant of a surface once near sea level (Figure 2.2). It is very difficult to speak of plains in modern geomorphology as these relief elements carry the drawback of being linked to denudation chronology. Here I am not referring to peneplains nor to correlation of heights even if done with a computer, but to planation surfaces whose relief-forming mechanisms have been studied and compared with those in tropical countries. Budel (1965 p.77f) names 15 different criteria for "Altflachen"(ancient plains). By actualistic reasoning we can conclude that high-lying planar relief elements of a certain size (at least a football field) recurring in several places are witnesses of a former extended plain.

Quite often this is supported by correlative weathering relics on the relict plains like the Augensteine in the Alps or by sediments in the foreland which may prove intensive weathering of the uplifted source area. Furthermore we can analyse additional relief elements of tropical origin like broad valleys, planation surfaces, intramontane basins, or even glacial stairways for which Bakker (1965) pointed out a tropical heritage. If there is a flight of different planation levels, a Rumpftreppe, this might suggest intermittent uplift. Perhaps planar remnants at different heights which often make the reasoning for an extended planation surface difficult, are evidence of a more continuous uplift.

In the first phases of uplift documented by relict plains the movement may be rather slow as planational lowering keeps pace with tectonically induced uplift. The top or even the upper half of anticlines may be cut off by this planational lowering. The height of mountains geologically deduced from structure is not realistic. It implies that the total amount of uplift took place before the beginning of erosion.

If we find dated sediments in the uplifted mountains or if we can apply modern dating methods we have a time marker for the uplift. But certainly we cannot infer linear uplift rates. There are times of faster and slower uplift and most likely of quiet scenes or even of intermittent subsidence as is shown by the columns of Pozzuoli (Schwarzbach 1976).

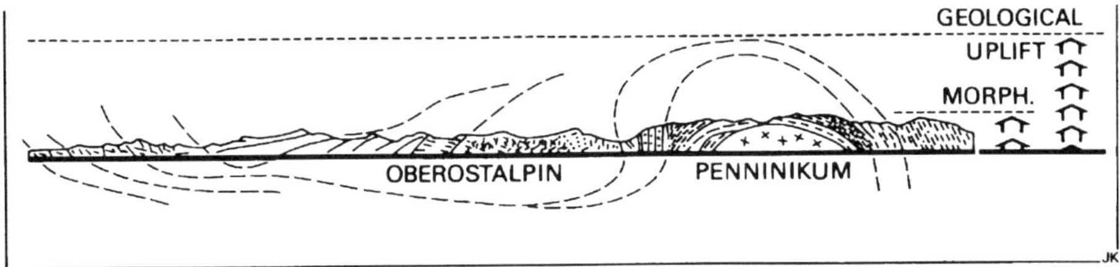

FIGURE 2.2 The height of uplift above sea level is called "morphological uplift", the extrapolation of geological strata assumes a height of mountains which never existed because in the first phases of uplift planational processes keep the land near sea level (Bremer 1989, p. 336). The diagram shows a schematic north–south section of the Alps

Denudation rates are also not constant either in time or in space. This applies not only at the short-term scale but also for the long-term average. After any major disturbance, whether climatically or tectonically caused, there may be a rapid response. This declines slowly asymptotically. Spatially there are areas which have not been lowered since Eocene time, like planation relicts in the Suebian Alb, which is proven by fossil karst fissures in the same area as recent valley floors (Kuppels 1981; Bremer, Liedtke and Spath 1989). Under tropical conditions bare rock surfaces are very resistant while neighbouring areas with soils are undergoing further weathering and erosion. This was called divergent weathering and erosion (Bremer 1981). It applies not only to inselbergs but also to escarpments.

Rates of uplift and erosion thus give us only a general idea about magnitudes involved. They do not explain landforms let alone geomorphological processes. To sum up: The second proposition is quite correct in so far as denudation and uplift are concurrent. The rates are individually different and must be derived by field investigations.

The third proposition of Brunsden (1990) is concerned with base level and the change of sea level. Certainly in Cretaceous time sea level was 200–400 m higher as may be deduced from extended Cretaceous deposits at this height in different continents. For an explanation of this higher sea level we might look at plate movements. With high spreading velocities ocean bottoms do not sink rapidly. The ocean bottoms are therefore not so deep, and the surface of the water is higher.

The assumption that landforms, especially plains, were related to these higher sea levels has not been verified by field observations. In small areas this is almost impossible to distinguish from uplift. I don't think we will ever believe in the continent-wide planation levels which were discussed in the 1930s and at the Washington Congress of the International Geographical Union in 1952.

We know of low sea levels during the Ice Age. They are connected with scour and fill in the lower parts of the rivers. These investigations have to take into account the gradient of the shelf. Lower sea levels may show at coral reefs, too.

In conclusion, we should look again into the question of changes of sea level in relation to geomorphology. At what time and spatial dimensions is the ultimate base-level relevant? We should think not only of base level as the ultimate control to transport. Moisture conditions at the weathering front are controlled by base level, too. This is most important in the tropics.

Recent discussion has brought about a wide array of theories, models and methods: from equilibrium as a general framework to the measurement of erosion on a small plot. We need paradigms not only for scientific discussion but for fundamental reasoning. All theories and paradigms have been questioned and mostly with good arguments. As an example Baker and Twidale (1991) may be cited. They try to convince geomorphologists in an almost poetical way after denying that theories and paradigms are true to nature. The trouble is, if there is little left in general rules or methods, everyone has to start from nothing. Their proposal to go into the field and find rational explanations there will eventually lead to the reinvention of climatic and climatogenetic geomorphology, to a repeating of the formulation of old theories.

Some of the misunderstanding in discussion about different theories stems from the neglect of a strict observance of scale in time and space (Bremer 1989; Slaymaker 1991, 1992). The second pitfall is the neglect of friction and inertia in the widest sense which is responsible for thresholds which may be tremendously large and relaxation times which may span millions of years. The third obstacle is the neglect of the paleo-character of many landforms.

We need theories, paradigms and systems, and constructive debate about them. Certainly Brunsden's (1990) paper is a clear and comprehensive basis for discussion. Most of his propositions can be verified by observations in nature or have been developed from empirical data. This allows for further refinement of theories and of changes in the boundary conditions. By constant checking back with nature we will eventually have a network of approaches which have a universally accepted core.

REFERENCES

Baker, V.R. and Twidale, C.R. 1991. The reenchantment of geomorphology. *Geormorphology*, **4**, 73–100.

Bakker, J.P. 1965. A forgotten factor in the interpretation of glacial stairways. *Z. Geomorph., N.F.*, **9**, 18–34.

Barsch, D. 1990. Geomorphology and geoecology. *Z. Geomorph., Suppl. Bd*, **79**, 39–49.

Bremer, H. 1959. Fluss erosion aner Oberer Weser. *Göttinger Geographisch Abhandlung*, **22**.

Bremer, H. 1981. Reliefformen und reliefbildende Prozesse in Sri Lanka. In Bremer et al. (1981), pp. 7–183.

Bremer, H. 1985. Randschwellen: a link between plate tectonics and geomorphology. *Z. Geomorph, Suppl. Bd.*, **54**, 11–21.

Bremer, H. 1989. *Allgemeine Geomorphologie*. Gebruder Borntraeger, Stuttgart.

Bremer, H., Schnutgen, A. and Spath, H. 1981. *Zur Morphogenese in den feuchten Tropen. Verwitterung und Reliefbildung am Beispiel von Sri Lanka. I Relief, Boden, Palaoklima.* Stuttgart.

Bremer, H., Liedtke, H. and Spath, H. 1989. *From the Alps to the sea. Field trip A. Geomorphology in Germany*. Geookotorium Darmstadt.

Brunsden, D. 1990. Tablets of stone: toward the Ten Commandments of geomorphology. *Z. Geomorph., Suppl.*, **79**, 1–37.

Brunsden, D. 1993. The persistence of landforms. *Z. Geomorph., Suppl.*, **93**, 13–28.

Budel, J. 1965. Die Relieftypen der Flachenspulzone Sud-Indiens am Ostabfall Dekans gegen Madras. *Coll. Geogr.*, **8**, Bonn.

Budel, J. 1982. *Climatic geomorphology* (translated by L. Fischer and D. Busche). Princeton University Press.

Godard, A. 1990. Rhythms in the geomorphological evolution and time–space related scale levels in the high latitudes. *Z. Geomorph., Suppl.*, **93**, 61–68.

Kennedy, B.A. 1992. Hutton to Horton: views of sequence, progression and equilibrium in geomorphology. *Geomorphology*, **5**, 231–250.

Kuppels, I. 1981. Die Karstspalten der Schwabischen Alb als Leiftformen fur die Morphogenese. *Kolner Geogr. Arb.*, **39**.

Schwarzbach, H. 1976. *Europaische statten geologisches Forschung*. Stuttgart.

Slaymaker, O. 1991. Mountain geomorphology: a theoretical framework for measurement programmes. *Catena*, **18**, 427–437.

Slaymaker, O. 1992. Ahnert's process-response models of denudation and the scale of dependence of sediment yield models — an attempted reconciliation. *Catena Suppl.*, **23**, 125–134.

Slaymaker, O. 1993. Holocene sediment budgets and environmental change in alpine environments. *Pre-Conference Field Excursion Guide A2, International Association of Geomorphologists, August 12–23, 1993.*

EXCEPTIO PROBAT REGULAM: DO EXCEPTIONAL FORMS TEST GENERAL THEORIES?

C.R. TWIDALE

SUMMARY

Though Brunsden's Ten Commandments are evidence of profound understanding of the factors that determine the shape of the Earth's surface, they are abstruse in expression and so specific in content as to preclude adjustment in the face of changing evidence and concepts. Two substantive questions arise. First, given the spatial and temporal complexities of the Earth's surface, is it possible to derive general statements or laws about its development? And, indeed, is it desirable to do so? Though a framework is useful for pedagogic purposes, it may take on an unjustified aura of respectability and immutability, and stultify rather than advance research and understanding. Second, if guidelines are sought, what factors which Brunsden overlooked or underplayed deserve mention? Water is so significant in several aspects of the denudational complex that it warrants specific consideration. Interactions between bedrock and shallow groundwaters produce etch forms which are widely developed and which are climatically and lithologically largely azonal. Subsurface weathering and flushing may contribute substantially to surface lowering. The survival of very old paleoforms in some instances carries significant implications for the distribution of water erosion in some (structural) conditions, as well as explaining several otherwise puzzling aspects of landscape development such as increasing relief amplitude, relief inversion, scarp steepening and developments in the piedmont zone.

INTRODUCTION

One of the primary aims of any scientific endeavour is to derive general laws, whatever name they are given (principles, models etc). To be useful, laws must be simple, clear, and comprehensible to all. Even if, like some of the biblical Ten Commandments, they are in practice difficult or impossible to observe in all circumstances, the best laws (again like the Ten Commandments) put otherwise unimagined ideas into people's heads; both the insights they represent and their shortcomings suggest modifications of or extensions to contemporary thought; they stimulate lateral thinking. To frame any general law entails treading a knife edge between the obvious, or apparently self-evident statement, and that which in an effort to be particular, becomes too specific and loses the flexibility necessary to achieve and maintain a measure of generality.

As has been made abundantly clear by such writers as Schumm (1985), as well as Brunsden (1990, p. 25, Proposition 9), landscape development is complex in both the spatial and temporal senses. General laws may be inappropriate not only at this but at any stage of the development of the discipline. Simple reductionism does not begin to meet our needs; an holistic and chronological approach is essential if time and space are to be integrated.

Thankfully, it will never be possible to reduce geomorphological analysis to a flow sheet. As Kant pointed out 200 years ago, truth is not reality but rather reality as perceived by the human mind, and such is the ingenuity of the human mind that controversy and debate will always be the rule rather than the exception. Concepts and theories can never be regarded as established, if only because the data on which they are based are inevitably flawed or incomplete, and because concepts are modified or replaced. No matter how much we posture, how much we deceive and indulge ourselves, there are no certainties, only probabilities: to be uncertain may be uncomfortable but to be certain is to court disaster. At best, a geomorphological "certainty" can be no more than reasonable probability. In any case, too rigid and manifestly plausible a set of laws may stultify ultimate progress by inhibiting critical thought.

On the other hand, guidelines that are undesirable or unnecessary for the experienced researcher, may have pedagogic value, focusing the attention of students on critical factors or concepts, and providing a framework within which complex data and ideas may be presented. And for the research worker, too, it is useful from time to time to take stock. The search for universal truths must not, however, be allowed to consume too much time or intellectual energy, and must not divert attention from conversations with the Earth — and other planets. We must not be seduced by stimulating, clever debates on the nature of geomorphology, similar to those on the nature, place and function of geography in the years following World War II (see Hartshorne 1939), debates which, though far from sterile, proved, as shown by subsequent developments (Baker and Twidale 1991), to be ineffective.

Several sets of guidelines, spanning both what to do and what not to do, have been suggested by previous workers. For example King's (1953) 50 canons contain much of value, yet some seem strained and contrived. Thornbury's (1954) nine fundamental concepts cannot be adversely criticized on the grounds of their being prolix but, as with King's canons, they inevitably reflect the background of the author. Schumm (1985) astutely points to problems inherent in attempts to formulate general geomorphological statements, and Brunsden (1990), in arriving at ten propositions, reveals profound insights into the dynamics of our world. His propositions reflect combinations and elaborations of Davis' (1899, 1909) well-known trinity of Structure, Process and Time, in terms of modern concepts. Yet they are so complex and esoteric in expression as to be difficult to relate to reality, and so specific as to be vulnerable in the face of new developments. They are of limited usefulness in the field.

Brunsden's Ten Commandments can be considered *seriatim* and defects, and exceptions, general or local, are readily identified. Consider for instance Brunsden's (Proposition 3, p. 11) emphasis on sea-level as a limit and control on landscape evolution. As a general statement pertaining to base-level it makes good sense, but base-level is, of course, not necessarily synonymous with sea-level. Huge areas of Australia are endoreic, and of them substantial areas are not areic but are drained by rivers that (occasionally) run to interior basins. For example the catchment focused on Lake Eyre occupies an area of some 1.3 million km^2. The development of such endoreic drainage basins owes little or nothing to sea-level but is, in the sense of base-level, dependent on the tectonic evolution of the region (see e.g. Wopfner and Twidale 1967). Moreover, Lake Eyre stands some 15 m below sea-level and for this reason it is likely that its catchment will expand at the expense of those of exoreic streams rather than vice versa. Also, even steep gradient rivers like those flowing to the Pacific coast are not extending inland as rapidly as might be anticipated: some have headed back only 100 km in the last 60 Ma (Taylor et al. 1985).

Such criticisms could be made of any but the most general and obvious propositions. They are of little moment. Of greater concern is the language in which Brunsden's propositions are couched. It seems unnecessarily obscure, and jargonistic. Moreover, the use of the term *tectonic* and its derivatives in the sense in which it is used by Brunsden, is surely questionable. *Tectonic* pertains to crustal processes, and tectonic forms are those due to active crustal effects such as volcanism or faulting. Forms due to exploitation of weaknesses in a passive crust by external agencies are structural, *sensu stricto*, and structural and tectonic together are conventionally subsumed under the term structural, *sensu lato*.

But rather than dwell on perceived shortcomings, it is more constructive to explore alternative factors and to consider certain of Brunsden's propositions in that context. To this end I shall develop some general remarks concerning models of landscape evolution in the context of the reality of ancient landscapes, and of etching or subsurface initiation.

Read (1957, p. xi) remarked that "... the interpretation of field-work is often a strictly personal affair, depending on the observer's character, training and experience". And certainly background and experience must be reflected in the derivation of general statements concerning factors responsible for landscape development. My experience has been mainly in ancient cratonic areas, and in tropical and subtropical lands. Though I have from time to time become especially interested in particular facets of the landscape, I have, for my sins, perforce remained essentially a generalist; for it is difficult to overlook any facet or factor if a stratigraphic or chronological approach is seen as fundamental (Baker and Twidale 1991).

Let me at once declare some of my prejudices, for I must confess that the Davisian trinity appeals to me, as it did to Hills (1961) and many others, and for good reasons (see e.g. Higgins 1976), as simple, straightforward and pertinent. That it was enunciated almost a century ago does not render it invalid. Nor does its authorship rule it out of court. It is sufficiently general to allow of flexibility and adjustment to new evidence and concepts, as it is to the almost infinite variations found in nature over time; for truly, the most common quality of landscape is diversity. The Davisian framework also enjoys the enormous advantage of being immediately useful both in the field, and as a teaching framework.

In Australia most ancient epigene surfaces are, in large measure, of etch type. Some retain patches or pockets of the original regolith especially where the latter is indurated, but over wide areas it has been stripped, though the essential morphology of the epigene form is retained (see e.g. Hills 1975, p. 300). Similarly, those regoliths that survive have developed throughout their exposure though their original character can usually be estimated (e.g. McFarlane 1986). Some regoliths have been stripped after dissection, but some, associated with exhumed features, were removed during the marine transgressions responsible for burial; only aeolian or volcanic covers are conducive to the preservation of regoliths. Old and etch forms together occupy substantial areas of the continental landscapes, particularly in cratonic regions, but also in the older orogenic belts. Old paleoforms and etch features together do not occupy the greater part of the continents, though they are widespread and account for substantial areas. What implications do these exceptions carry for general theories of landscape development?

ETCH FORMS AND THEIR IMPLICATIONS

The essentials of etching were appreciated at least 200 years ago, when Hassenfratz (1791) realized that some granite boulders form beneath the surface as a result of differential subsurface weathering. A little later, Logan (1849, 1851) demonstrated that some flutings in granite are initiated beneath the soil cover. Falconer (1911) perceptively deduced that some inselbergs are of subsurface origin, and Jutson (1914) reached the same conclusion concerning some of the high plains of Western Australia, but it was Wayland (1934) and Willis (1936) who together applied the terms *etch* and *etching* to the process of shaping of the rock surface at the base of the regolith, at the weathering front, that is, the contact between the regolith and intrinsically fresh country rock (Mabbutt 1961).

The weathering front develops a relief both in gross and in detail as a result of the exploitation by groundwaters of structural weakness in the bedrock. Such bedrock forms are, after exposure, known as etch forms. As Lewis (1955) pointed out, etching may occur in two distinct phases or essentially a single operation where erosion keeps pace with weathering; and Büdel (1957) suggested that even in two-phase situations erosion may simultaneously operate at two levels, reducing the land surface and exposing the weathering front: *die Doppelten Einebnungsflächen* or double planation.

Such etch forms have two ages, one indicating the phase of subsurface preparation, the other the period of exposure of the weathering front. For this reason, etch landforms and landscapes are commonly referred to as being of two-stage type. In reality they are multistage, with magmatic, thermal and tectonic events leaving their imprint on the etch landscape. Several phases of evolution have already been recognized in some landscapes (Twidale and Vidal Romani 1994). Take for example the bornhardts of the Gawler Ranges, in the arid interior of South Australia. They are etch forms, having been developed by subsurface weathering in Jurassic or earlier times. They were exposed as landforms during the Early Cretaceous (Neocomian–Aptian), for detritus, including corestones, evacuated during this stripping has been located in dated sedimentary sequences to the north and northeast, in the Eromanga Basin. The present valley floors and piedmont plains were formed and duricrusted by, or during, the Eocene: though the bornhardts are of etch type, they are nevertheless ancient. But the subsurface weathering responsible for shaping the bornhardts exploited orthogonal fracture systems that were developed by 1050 Ma. Thus three stages in the development of the bornhardts can be identified and dated: fracture formation, differential fracture-controlled subsurface weathering, and erosion of the regolith to expose the differentiated weathering front. To these might be added a fourth, the initial stage, namely the extrusion of the ash-flow tuffs from which the massif constructed almost 1.6Ga (Campbell and Twidale 1991).

Etch forms are lithologically largely azonal, and, because the groundwaters that are largely responsible for etching are essentially ubiquitous, they are also climatically azonal (Twidale 1990). Their rate of development may have varied according to atmospheric climate and rock type, but the end results are the same. This is consistent with Brunsden's Proposition 4 (1990, p. 14), though coincidentally rather than intentionally, but is at odds with Proposition 6 (1990, p. 18) with its implication of strong climatic influence, if not control. Twidale and Lageat (1994), in some respects elaborating on comments by Tricart and Cailleux (1972), have called in to question certain aspects of the morphogenetic concept, pointing out that although climatic accidents (arid, glacial, nival) leave their imprint on the landscape, a multitude of other regions together comprising about 50% of the continental areas of the Earth, owe their morphology to structural factors, to etching and to the work of exogenic processes that transgress climatic boundaries. Many familiar landforms, major and minor, are common to several of our conventionally defined climatic regions.

Subsurface weathering and flushing may explain otherwise puzzling landscape features. In some areas, rolling plains are developed but there are few surface streams or stream channels. The Sturt Plateau, in the Northern Territory, is a lateritized plain developed in Early–Middle Tertiary times on Cretaceous siltstone and sandstone. There are a few widely spaced shallow channels, but no close network of stream valleys. To the west and southwest of Lake Eyre and graded to that regional base-level, the extensive, broadly rolling, Glendambo plains developed on Cretaceous argillites show little evidence of surface drainage. What agencies are responsible for shaping these planate landscapes?

Overland wash undoubtedly occurs and assists in the shaping of the surface, but many years ago Trendall (1962) suggested that chemical weathering, loss of volume, settling and compaction, contribute substantially to the lowering of plains. Though there have been objections (e.g. Ollier 1971), it seems feasible that, given time, chemical alteration can reduce and lower regions underlain by any type of bedrock. The concept is consistent with the paucity of concentrated lines of surface drainage, though widely spaced underground flows may well be developed and lead to pronounced lowering in essentially linear depressions or valleys.

The available data suggest that, on average, rivers flowing to the oceans transport almost 4 billion tonnes of materials in solution per annum (Livingstone 1963). Admittedly this is merely of the same order of magnitude as the volume of detritus moved from quarries, mines and cuttings every year but is nevertheless impressive, especially as it takes no account of transport in groundwater seepages or in endoreic basins. Nor do the figures include any fines flushed from the system (Ruxton 1958). Carbonates are quantitatively dominant in the list of solutes but silica and sulphates are also prominent. The data suggest that Australia is at present little affected by such processes, with only 15.5 metric tonnes of solutes lost for every square kilometre, compared with some 47 tonnes/km in Europe and over 36 for each of North (including Central) America and Asia (Livingstone 1963); but with a third of the continent served by interior drainage and with the data pertaining to present aridity rather than past humid conditions (see e.g. Kemp 1978) the reality is rather different, especially if the time available for such processes is borne in mind.

Weathering may convert fresh, intractable material to a size susceptible to transport, so that waves of erosion may develop as a result of alteration of the country rock and have no connection with base-level shifts. Long periods of weathering produce intense alteration and duricrusts (see below). Some minerals are concentrated within the regolith and later, after dissection, give rise to structural plains and plateaus remote from regional or ultimate base-level (e.g. Wright 1963). Duricrusts are conducive to the development of relief inversions on a regional scale. Within and marginal to the Hamersley Ranges of the northwest of Western Australia, the Robe River Pisolite caps and protects the many spectacular elongate sinuous mesas that represent Eocene valley floors (Twidale et al. 1985) and many silcrete and travertine valley-floor accumulations take a similar form.

Acting rapidly on susceptible bedrock like limestone, etching can explain the astonishing flatness of large parts of the Nullarbor Plain (Twidale 1990). On the other hand, long continued etching offers an explanation for the remarkable flatness of planation surfaces like that eroded in granite and gneiss around Meekatharra in the arid interior of Western Australia, and that cut in granite, gneiss and sandstone in southern

Namibia and northern Namaqualand (Cape Province, South Africa): it also accounts for the remarkably small plan size of residuals relative to the extent of the surrounding plains, in many shield inselberg landscapes. It has been argued with respect to bornhardts and inselberg landscapes that they have a common origin in the humid warm environments of the earlier Cainozoic; and this may well apply to many, though by no means all, examples. Bornhardts and inselbergs of etch origin have developed under various climatic conditions, in various rock types and at various times ranging back to the Archean. The common factor is a subsurface initiation.

Etch forms are as commonplace in glaciated as in tropical lands, for some glaciers have acted as bulldozers and have essentially stripped away the pre-existing regolith (Boyé 1950; Twidale 1990). Many of the glaciated valleys of the Yosemite of the central Sierra Nevada in California, for example, and glaciated rocky high plains like those of central Labrador, may well be largely of etch type.

Thus shallow groundwaters have produced several forms, major and minor, that are familiar features of the landscape, and the implications of which (e.g. azonality) are either at odds with, or are unaccounted for, in well-known models of landscape development.

ANCIENT SURFACES AND THEIR IMPLICATIONS

(a) The age of "old" surfaces and forms

Many landforms are much older than hitherto acknowledged or thought possible (e.g. Thornbury 1954; Brown 1981). Some are exhumed, and in Australia are of ages ranging from Archean to Pleistocene (for references see Twidale and Campbell 1988). In other continents exhumed forms include the classical inselberg landscapes described from what is now Tanzania by Bornhardt (1900; see Willis 1936). Some of them predate the rifting and drifting of Gondwana elements and many are of Late Jurassic or Early Cretaceous age, i.e. 60–160 Ma. Many remain preserved in unconformity and, overall, the survival of exhumed surfaces and forms poses few problems, though the survival of even minor karst forms such as solution hollows, and *Kluftkarren*, which manifestly predate burial, is surprising (e.g. Twidale 1984).

The occurrence of very old paleoforms in Australia is documented elsewhere (e.g. Twidale 1976, 1994; Twidale and Campbell 1988), but the island continent is not unique in the antiquity of some of its landscapes, as was demonstrated long ago by Falconer (1911) and King and his predecessors in Africa (see e.g. 1962 for references and see also Michel 1978; Partridge and Maud 1987), and other investigators working in India (Demangeot 1978) and on the Guyana Shield (Briceño and Schubert 1990). Surfaces of similar antiquity may well be preserved elsewhere, in cratonic and orogenic elements of the former Pangaea. For instance, not only have old exhumed surfaces been identified in glaciated areas of Greenland, Canada and Fennoscandia (e.g. Ambrose 1964; Fogelberg 1985), but the summit high plain of the central Sierra Nevada in California may well be of later Mesozoic age (Curtis et al. 1958) and epigene-etch surfaces may also be represented in complex composite features like the *Gipfelflur*. The problem here is that intense dissection has caused the destruction of all or most of the relevant evidence.

According to the well-known models of landscape evolution these ancient paleoforms ought to have been obliterated long ago. The time taken for base-levelling varies according to the character of the land mass, and particularly its areal extent and the criteria used. For example, using fission track dates, Pitman and Golovchenko (1991) have calculated that the base-levelling of major area has occupied periods as great as 108 Ma. On the other hand, using stratigraphic evidence, and calculations based on rates of erosion, such workers as Schumm (1963) have suggested that provided there were no major movements of base-level (resulting from tectonism, sea-level shifts or both), but taking account of isostatic adjustments consequent on erosional unloading, a small region like New Zealand would be base-levelled in some 11 Ma, whereas 33 Ma would elapse before an area like the continental United States would be reduced to base-level. The estimates derived by this second group for the duration of a geomorphic cycle are thus an order of magnitude different from those based on geophysical criteria; but they seem more in keeping with stratigraphy.

(b) Some implications of antiquity

However, antiquity brings clarification as well as problems, for some of the difficulties associated with the conventional models of landscape development are reduced or eliminated if the implications of time are considered. Time allows deep and intensely weathered regoliths to evolve beneath surfaces of low relief. It allows the reduction of resistant masses. It permits the development of duricrusts of various types and allows organic evolution to take effect. And these changes in turn may induce changes in the mode of landscape development. For example, given time, slopes become steeper and steeper until they attain the maximum inclination commensurate with stability. Evidence of such regrading of slopes is widespread in many parts of Australia, with scarp foot valleys and false cuestas widely developed (e.g. Twidale and Milnes 1983).

A long time perspective also directs attention to the possible impacts of changes effected during the course of a phase of denudation. Environments are constantly changing, and earlier events influence the impact of later, so that rather than being cyclic in character, time ought to be considered spiral, with landscapes tending to go full circle but returning, as it were, at a different level or in a different condition. Though cyclic development is not denied for some areas — the evidence of phases of landscape reduction between major tectonic events seems overwhelming — the suggestion that the end product of cyclic development is the same or even closely similar to the original surface is likely to be misleading. Factors built into the cyclic schemes may imply that the nature of the land surface at the end of the cycle differs from the "original". Thus in northwest Queensland, the land surface of low relief was lateritized during the Early–Middle Tertiary (Twidale 1956). On dissection of the surface, the ferruginous duricrust initially induced scarp retreat, but where the resistant capping was either eliminated or where it had never developed, slope lowering probably produced the present extensive complex and polycyclic peneplain. Earlier weathering events have influenced later slope development and behaviour.

Also, erosion may uncover rocks that differ in their weathering and erosional characteristics from those previously exposed. A granitic batholith may be uncovered — or it may rise through the superincumbent rocks (see Ollier and Pain 1980) — with the result that granitic forms replace say, ridge and valley developed on a folded sedimentary sequence and this change may embrace relief inversion where granite is weathered more readily than, say, metasediments (e.g. Vogt 1953). Furthermore, river patterns may also become anomalous as the watercourses penetrate into structures that vary geometrically with depth (see e.g. Meyerhoff and Olmstead 1936; Oberlander 1965; Twidale, 1966).

(c) Organic evolution

Organic evolution proceeds, progressively or regressively, gradually or spasmodically, but relentlessly, and during the time elapsed between beginning and end of a cycle or episode, new organic agencies may be at work. Cloud (1980, p.16) has stated "... after the successful establishment of terrestrial life, Earth surface processes could never be the same again." Given the great age of Gondwana elements it is reasonable to suggest that the biotic environment varied in successive phases of landscape development.

The impact of vegetation, and especially trees, on landform development is well illustrated by the effects of anthropogenic clearance, namely in the development of an epicycle of accelerated erosion. As Russell (1958) remarked of the Cainozoic land surface, it has never before been so well armoured. Again, eucalypts produce a litter rich in polyphenols that are conducive to iron solution, and they may have contributed materially to the formation of dolines in the laterite of the Sturt Plateau, Northern Territory (Twidale 1987). Silcrete is, as far as is presently known, a relic feature: its contemporary development has not so far been demonstrated. It may be that we have not yet examined the right vegetational niches (e.g. the tropical rain forest) but it may be that, as Lovering (1959) suggested, silcrete reflects the presence of plants capable of silica accumulation, plants like ferns, which were dominant in various regions during the Tertiary and Early Pleistocene. Both mineral carapaces and vegetational cover must have retarded erosion and planation.

The effects of bacteria are less readily assessed for little is known of the extent, purpose and effects of their penetration of various mineralogical niches (but see Trudinger and Swaine 1979). Bacteria have existed, however, for at least 3 Ga and there are hints that their impacts are substantial (see e.g. McFarlane and Heydeman 1984). Similarly, the effects of worms, especially earthworms, and termites in the geological context are difficult to assess and the same comments apply to various marine organisms such as the chitons, with their magnetite teeth, but their impact must be considerable. Many impacts may have been destructive, as for example by enhancing and increasing the rate of weathering, but others may have caused cementation and retardation of landform evolution.

(d) Implications of survival

Several factors commonly combine to permit survival. First, many of the ancient paleosurfaces are underlain by and are preserved in resistant bedrock: Banded Iron Formations (Hamersley Surface), ferruginous caps (Robe River Pisolite, Gulfs region of South Australia), well-drained silicic volcanic rocks (Gawler Ranges), quartzite (Kakadu, MacDonnell Ranges, southern Flinders Ranges, Isa Highlands), or granite, which when high and dry is stable. But some old surfaces are preserved in, for example, weak siltstone, as in the central Flinders Ranges; and here it may be, as in granitic terrains and in Ayers Rock, that compression due either to folding or to shearing is a key factor causing fractures to be tight and preventing significant percolation of water.

Second, and as emphasized in relation to etching, water is critical to weathering and if it can either be prevented from entering the rock mass or allowed freely to drain through, the chances of alteration are reduced (Barton 1916; Twidale and Campbell 1993). Thus blocks that stand high may also be dry and stable. In Australia many of the uplands carry remnants of paleosurfaces, and they have been repeatedly uplifted, most recently as a result of isostatic adjustments following the Early Cretaceous transgressions (Twidale and Campbell 1992; Twidale 1994). In southern Africa, too, the old land surfaces stand high in the relief, either as a result of isostatic uplift following erosional unloading or as a result of hot-spot activity in the underlying crust. In detail, various minor features such as rock doughnuts and rock levees have been attributed to contrasts in rates of weathering related to wet and dry sites (Twidale 1988).

An indication of the importance of moisture in landscape evolution is provided by the surface of the Earth's moon. It consists of impact craters and volcanic outpourings with depressions partly filled by a regolith consisting of fragments and dust produced by impacts. The surface is very ancient, with some sectors 4 Ga old, but in the absence of moisture the volcanic and impact landscape remains unchanged save for the destructive effects of further impacts (e.g. Wilhelm 1987).

Third, though scarp retreat and peneplanation appear to have operated in some areas, the former in regions of caprock, the latter in areas of weak outcrops, the field evidence in uplifted blocks that include resistant materials points to unequal activity with deep gorges or valleys separated by stable divides which are remnants of ancient surfaces. In other words, though attack may be intrinsically constant, it may also be localized.

Horton (1945) suggested that there is a region of nil erosion in the headwater zone of rivers and streams. Earlier Knopf (1924) had suggested that ancient parts of the Appalachians had survived because they were beyond the reach of stream erosion, and Crickmay (1932, 1976) took this as the basis for his concept of unequal activity, which stands in marked contrast with the widespread activity of streams and wash postulated by earlier workers (e.g. Davis 1909, pp. 266–267), and which is also implied in rates of erosion derived from the measurement of volumes of sediment derived from a particular catchment. The essence of Crickmay's thesis is that though, where base-level permits, streams erode intensively and rapidly in and near their channels, their divides may be lowered so slowly as to be essentially stable. The operation of the mechanism may be enhanced where the affected blocks have been uplifted either tectonically or isostatically (see e.g. Twidale and Campbell 1992). It leads to increased relief amplitude (Twidale 1991). The suggestion provides an explanation for those many areas where deep narrow valleys are incised below smooth surfaces or plains, and in lithological conditions that vary from fresh granite (Augrabies Gorge, Namibia; Herbert Gorge, northeast Queensland), quartzite (many rivers in the Cape Fold Belt; Edeowie Gorge, Wilpena Pound, Flinders Ranges; the Murchison in its gorge section, Carnarvon Basin, Western Australia), and basalt (Zambesi Gorge below the Victoria Falls; Taos River, New Mexico, many examples from north Queensland), to alluvium and regolithic materials.

Fourth, reinforcement or feedback mechanisms are a well-known feature of landform development at all scales. Some are negative, like the limitations inherent in the width of shore platforms because of the loss of energy due to friction as waves pass through shallow waters. Many reinforcement effects, however, are positive. For example, water penetration of rocks, as indicated by measured porosity and permeability, increases dramatically once the initial alteration of one or more constituents has been achieved. All drainage systems are examples of positive feedback systems but incised rivers especially so.

Thus time allows deep and intense weathering, with all that that implies for various facets of landscape development. It implies possible repercussive changes in slope behaviour, and biotic changes of significance to landscape development. Time allows conditions conducive to increasing relief amplitude, scarp steepening and associated piedmont developments, and relief inversion. Landform developments in the context of time measured in scores, even a few hundreds, of millions of years cannot be neglected.

CONCLUSION

For some workers, geomorphology is essentially chronological in character, and for those of this persuasion, denudation chronology, or the specific interplay of exogenetic and endogenetic forces and processes, is the key to understanding the morphology of landscape. Time is an essential element in geomorphological analysis. It allows many processes to take effect and in particular is conducive to etching. Groundwaters are ubiquitous and as a result etch forms are widely developed. In some areas water erosion is restricted and not general, as has been supposed. As a result old surfaces have survived for scores of millions of years. "... your water is a sore decayer", but appreciation of its localized activities beneath and at the surface helps explain many otherwise puzzling landscape features. Water is the critical factor in landform evolution on Earth. It surely warrants specific mention in any general guidelines such as those formulated by Brunsden.

REFERENCES

Ambrose, J.W. 1964. Exhumed palaeoplains of the Precambrian Shield of North America. *Amer. J. Sci.*, **262**, 817–857.

Baker, V.R. and Twidale, C.R. 1991. The reenchantment of geomorphology. *Geomorph*, **4**, 73–100.

Barton, D.C. 1916. Notes on the disintegration of granite in Egypt. *J. Geol.*, **24**, 382–393.

Bornhardt, W. 1900. *Zur Oberflächengestaltung und Geologie Deutsch Ostafrikas*. Reimer, Berlin, 595 pp.

Boyé, M. 1950. *Glaciaire et Périglaciaire de l'Ata Sund, Nord-oriental Groenland*. Herrman, Paris.

Briceño, H.O. and Schubert, C. 1990. Geomorphology of the Gran Sabana, Guyana Shield, Venezuela. *Geomorph.*, **3**, 125–141.

Brown, E.H. 1981. Historical geomorphology — principles and practice. *Z. Geomorph. Suppl.-Bd.*, **36**, 9–15.

Brunsden, D. 1990. Tablets of stone: towards the Ten Commandments of geomorphology. *Z. Geomorph. Suppl.-Bd.*, **79**, 1–37.

Budel, J. 1957. Die 'doppelten Einebnungsflächen' in den feuchten Tropen. *Z. Geormorph.*, **1**, 201–228.

Campbell, E.M. and Twidale, C.R. 1991. The evolution of bornhardts in silicic volcanic rocks, Gawler Ranges, South Australia. *Aust. J. Earth Sci.*, **38**, 79–93.

Cloud, P. 1980. Early biogeochemical systems. In Trudinger, P.A. and Walter, M.R. (eds), *Biogeochemistry of Ancient and Modern Environments*. Australian Academy of Science, Canberra, pp. 7–27.

Crickmay, C.H. 1932. The significance of the physiography of the Cypress Hills. *Canadian Field Naturalist*, **46**, 185–186.
Crickmay, C.H. 1976. The hypothesis of unequal activity. In Melhorn, W.N. and Flemal, R.C. (eds), *Theories of Landform Development*. SUNY, Binghamton, New York, pp. 102–109.
Curtis, G.H., Evernden, J.F. and Lipson, J. 1958. Age determination of some granite rocks in California by the potassium–argon method. *Division of Mines California, Spec. Rept*, **54**.
Davis, W.M. 1899. The geographical cycle. *Geogr. J.*, **14**, 481–504.
Davis, W.M. 1909. The geographical cycle. In *Geographical Essays*. Dover, Boston, pp. 249–278.
Demangeot, J. 1978. Les reliefs cuirassés de l'Inde du Sud. *Trav. Doc. Géogr. Trop.*, **33**, 97–111.
Falconer, J.D. 1911. *Geology and Geography of Northern Nigeria*. Macmillan, London, 295 pp.
Fogelberg, P. (ed.) 1985. Preglacial weathering and planation. *Fennia*, **163**, 283–383.
Hartshorne, R. 1939. *The Nature of Geography*. Association of American Geographers, Lancaster, Pennsylvania, 482 pp.
Hassenfratz, J-H. 1791. Sur l'arrangement de plusieurs gros blocs de différentes pierres que l'on observe dans les montagnes. *Ann. Chim.*, **11**, 95–107.
Higgins, C.G. 1976. Theories of landform development. A perspective. In Melhorn, W.N. and Flemal, R.C. (eds), *Theories of Landform Development*. SUNY Binghamton, New York, pp. 1–28.
Hills, E.S. 1961. Morphotectonics and the geomorphological sciences with special reference to Australia. *Quart. J. Geol. Soc. London*, **117**, 77–89.
Hills, E.S. 1975. *Physiography of Victoria*. Whitcombe and Tombs, Melbourne, 373 pp.
Horton, R.E. 1945. Erosional development of streams and their drainage basins. *Geol. Soc. Amer. Bull.*, **56**, 275–370.
Jutson, J.T. 1914. An outline of the physiographical geology (physiography) of Western Australia. *Geol. Surv. W. Aust. Bull.*, **61**.
Kemp, E.M. 1978. Tertiary climatic evolution and vegetation history in the southeast Indian Ocean region. *Palaeogeogr. Palaeoclim. Paleoecol.*, **24**, 169–208.
King, L.C. 1953. Canons of landscape evolution. *Geol. Soc. Amer. Bull.*, **64**, 721–752.
King, L.C. 1962. *Morphology of the Earth*. Oliver and Boyd, Edinburgh, 699 pp.
Knopf, E.B. 1924. Correlation of residual erosion surfaces in the eastern Appalachians. *Geol. Soc. Amer. Bull.*, **35**, 633–668.
Lewis, W.V. 1955. pp. 483–484 in Discussion (pp. 482–486) of Linton, D.L. The problem of tors. *Geogr. J.*, **121**, 470–487.
Livingstone, D. 1963. Chemical composition of rivers and lakes. Chapter G in Data of Geochemistry, 6th edition. *U.S. Geol. Surv. Prof. Pap.*, **440-G**.
Logan, J.R. 1849. The rocks of Palo Ubin. *Genootschap Kunsten Wetenschappen (Batavia)*, **22**, 3–43.
Logan, J.R. 1851. Notices of the geology of the Straits of Singapore. *Quart. J. Geol. Soc. London*, **7**, 310–344.
Lovering, T.S. 1959. Geological significance of accumulator plants in rock weathering. *Geol. Soc. Amer. Bull.*, **7**, 781–800.
Mabbutt, J.A. 1961. 'Basal surface' or 'weathering' front. *Proc. Geologists' Assoc. London*, **72**, 357–358.
McFarlane, M.J. 1986. Geomorphological analysis of laterites and its role in prospecting. *Geol. Surv. India Mem.*, **120**, 29–40.
McFarlane, M.J. and Heydeman, M.T. 1984. Some aspects of kaolinite dissolution by a laterite-indigenous micro-organism. *Geo-Eco-Trop.*, **8**, 73–91.
Michel, P. 1978. Cuirasses bauxitiques et ferrugineuses d'Afrique occidentale. Aperçu chronologique. *Trav. Doc. Géogr. Trop.*, **33**, 11–32.
Meyerhoff, H.A. and Olmstead, E.W. 1936. The origins of Appalachian drainage. *Amer. J. Sci.*, **36**, 21–42.
Oberlander, T. 1965. The Zagros Streams. *Syracuse Geographical Series*, **1**.
Ollier, C.D. 1971. *Weathering*. Oliver and Boyd, Edinburgh, 304 pp.
Ollier, C.D. and Pain, C.F. 1980. Actively rising surficial gneiss domes in Papua New Guinea. *J. Geol. Soc. Aust.*, **27**, 33–44.

Partridge, T.C. and Maud, R.R. 1987. Geomorphic evolution of southern Africa since the Mesozoic. *Trans. Geol. Soc. S. Africa*, **90**, 179–208.
Pitman, W.C. and Golovchenko, X. 1991. The effect of sea level changes on the morphology of mountain belts. *J. Geophys. Research*, **98**, 6879–6891.
Read, H.H. 1957. *The Granite Controversy*. Murby, London, 430 pp.
Russell, R.J. 1958. Geological geomorphology. *Geol. Soc. Amer. Bull.*, **69**, 1–22.
Ruxton, B.P. 1958. Weathering and subsurface erosion in granite at the piedmont angle, Balos, Sudan. *Geol. Mag.*, **45**, 353–377.
Schumm, S.A. 1963. Disparity between present rates of denudation and orogeny. *U.S. Geol. Surv. Prof. Pap.*, **454**.
Schumm, S.A. 1985. Explanation and extrapolation in geomorphology: seven reasons for geologic uncertainty. *Trans. Jap. Geomorph. Union*, **6**, 1–18.
Taylor, G., Taylor, G.R., Bink, M., Foudoulis, C., Gordon, I., Hestrom, J., Minello, J. and Whippy, F. 1985. Pre-basaltic topography of the northern Monaro and its implications. *Aust. J. Earth Sci.*, **332**, 65–71.
Thornbury, W.D. 1954. *Principles of Geomorphology*. Wiley, New York, 618 pp.
Trendall, A.F. 1962. The formation of 'apparent peneplains' by a process of combined lateritisation and surface wash. *Z. Geomorph.*, **6**, 183–197.
Tricart, J. and Cailleux, A. 1972. *Introduction to Climatic Geomorphology* (translated by C.J. Kiewiet de Jonge). Longman, London, 295 pp.
Trudinger, P.A. and Swaine, D.J. (eds) 1979. *Biogeochemical Cyclicity of Mineral Forming Elements*. Elsevier, Amsterdam, 612 pp.
Twidale, C.R. 1956. Chronology of denudation in northwest Queensland. *Geol. Soc. Amer. Bull.*, **67**, 867–882.
Twidale, C.R. 1966. Chronology of denudation in the southern Flinders Ranges, South Australia. *Trans. R. Soc. Aust.*, **90**, 3–28.
Twidale, C.R. 1976. On the survival of palaeoforms. *Amer. J. Sci.*, **276**, 77–94.
Twidale, C.R. 1984. The enigma of the Tindal Plain, Northern Territory. *Trans. R. Soc. S. Aust.*, **108**, 95–103.
Twidale, C.R. 1987. Sinkholes (dolines) in lateritised sediments, western Sturt Plateau, Northern Territory, Australia. *Geomorph.*, **1**, 33–52.
Twidale, C.R. 1988. Granite landscapes. In Moon, B.P. and Dardis, G.F. (eds), *The Geomorphology of Southern Africa*. Southern Book Publishers, Johannesburg, pp. 198–230.
Twidale, C.R. 1990. The origin and implications of some erosional landforms. *J. Geol.*, **98**, 343–364.
Twidale, C.R. 1991. A model of landscape evolution involving increased and increasing relief amplitude. *Z. Geomorph.*, **35**, 85–109.
Twidale, C.R. 1994. Gondwanan (Late Jurassic and Cretaceous) palaeosurfaces of the Australian Craton. *Palaeogeogr. Palaeoclim. Palaeoecol.*, **112**, 157–186.
Twidale, C.R. and Campbell, E.M. 1988. Ancient Australia. *GeoJ.*, **16**, 339–354.
Twidale, C.R. and Campbell, E.M. 1992. Geomorphological development of the eastern margin of the Australian Shield. *Earth Surf. Proc. Landforms*, **17**, 319–331.
Twidale, C.R. and Campbell, E.M. 1993. Fractures: a double edged sword. *Z. Geomorph.*
Twidale, C.R. and Lageat, Y. 1994. Climatic geomorphology: a critique. *Prog. Phys. Geogr.*, **18**, 319–334.
Twidale, C.R. and Milnes, A.R. 1983. Slope processes active late in arid scarp retreat. *Z. Geomorph.*, **27**, 343–361.
Twidale, C.R. and Vidal Romani, J.R. 1994. The multistage origin of etch forms. *Geomorph.*, **11**, 107–124.
Twidale, C.R., Horwitz, R.C. and Campbell, E.M. 1985. Hamersley landscapes of the northwest of Western Australia. *Rev. Géogr. Phys. Géol. Dynam.*, **26**, 173–186.
Vogt, J. 1953. Un probléme morphologique du bouclier canadien: le relief granitique. *Rev. Géomorph. Dynam.*, **4**, 85–95.
Wayland, E.J. 1934. Peneplains and some erosional landforms. *Geol. Surv. Uganda Ann. Rept. & Bull.*, **1**, 77–79.

Wilhelm, D. 1987. The geologic history of the Moon. *U.S. Geol. Surv. Prof. Pap.*, **1348**.

Willis, B. 1936. East African plateaus and rift valleys. *Studies in Comparative Seismology, Carnegie Institute (Washington, DC) Publication*, **470**.

Wopfner, H. and Twidale, C.R. 1967. Geomorphological history of the Lake Eyre basin. In Jennings, J.N. and Mabbutt, J.A. (eds), *Landform Studies from Australia and New Guinea.* Australian National University Press, Canberra, pp. 118–143.

Wright, R.L. 1963. Deep weathering and erosion surfaces in the Daly River basin, Northern Territory. *J. Geol. Soc. Aust.*, **10**, 151–164.

GRAFFITI ON THE WALL OF A GEOMORPHOLOGY LABORATORY

EIJU YATSU

SUMMARY

In general, propositions are subject to falsification and statements without falsifiability are called pseudopropositions. Commandments seem to be beyond falsification. Therefore the term "commandment"should not be replaced by the term "proposition", i.e. any elaborate proposition is not yet qualified to become a commandment as the goal of geomorphological research.

A special assignment allocated by the moderator is to focus my remarks on Propositions 5, 7, 8 and 9. A careful reading of these propositions reveals that Brunsden has succeeded skilfully in synthesizing the basic principles of geomorphology, surveying thoroughly existing geomorphological archives. His contribution should be highly appreciated. However, some questions and objections arise: (1) in discussing the perturbations to which landforms are continuously subject, he is too much occupied with explaining the importance of climatic changes (Proposition 5); (2) he defined the terms, catastrophe and geocatastrophe, but I would like to advance a proposal that the term catastrophe should be limited to the ancient discussion of catastrophism. A classification of geocatastrophe (Figure 2.3) seems dubious, and in addition, his usage of the term uniformitarianism in the context seems to undermine the principle of uniformity of nature (proposition 7); and (3) proposition 9 explains landscape stability very theoretically, but it is still conspicuously lacking in data obtained through scientific instrumentation.

The most serious problem surrounds the aims and objectives of the paper. Each proposition is certainly a summary of one aspect of the present situation of geomorphology. Therefore Brunsden (1990) is one of the best chronological archives and a most educational text for novices in this field, but it does not act as an incentive to progress in geomorphology. Advancement does not need mental obstacles but creative freedom.

Denys Brunsden has given us a fine analysis of geomorphological thought and its theoretical framework. For this part of his paper, I have nothing but admiration, and yet I shall not hesitate to build my argument on his address. To facilitate our discussion, here is a summary of my approach:

1. Comments on the title: Tablets of Stone.
2. Remarks on Propositions 5, 7, 8 and 9, and some general criticism.
3. Conclusion: A request addressed to President Brunsden.

These items will be briefly discussed in this order to avoid misunderstanding and confusion.

COMMENT ON THE TITLE: TABLETS OF STONE

For a person whose mother tongue is not English, it is, as one might expect, very difficult to answer a question such as "To what extent are some special divine or theological words in English allowed to be used metaphorically as ordinary language?" I find it difficult to imagine that words like "Tablets of Stone", or "Ten Commandments" could be used in the title of a scientific paper, even though these words might be chosen for fun. I felt it to be inappropriate and indeed I was somewhat shocked. I have already criticized his selection of the title of his paper elsewhere (Yatsu 1992).

Let me say one word more on "ten commandments and ten propositions". It seems to me observably false to substitute commandments in the subtitle with propositions in the text. In general, propositions are subject to falsification and statements without falsifiability are called pseudopropositions. Commandments seem to be beyond falsification. Therefore the term "commandment" in the subtitle cannot be logically replaced by the term "proposition", i.e. any elaborate proposition is not qualified yet to become a commandment as the goal of geomorphological research.

REMARKS ON PROPOSITIONS 5, 7, 8 AND 9, AND SOME GENERAL CRITICISM

A careful reading of Propositions 5, 7, 8 and 9 reveals that Brunsden has succeeded skilfully in synthesizing the basic principles of geomorphology, surveying thoroughly existing geomorphological archives. His contribution should be highly appreciated. However, some questions and objections arise.

In Proposition 5, he states that landforms are continually subject to perturbations which arise from changes in the environmental conditions of the system, and these impulses are episodic and complex in nature at all scales.

His claims may be all right. He asserts that landforms are the response to the controlling variables of tectonics, sea-level change, climate, climatic change and biotic activity over time, and accordingly they also respond to the changes, rhythms and threshold of earth. His analysis of geomorphological systems is exhaustive. The table and figures used in Proposition 5 are exclusively concerned with climatic change. Brunsden's interest in climatic geomorphology seems to be strongly supportive, whereas my view of the present state of climatic geomorphology will be exhibited later.

Proposition 7 defines geocatastrophe as follows: New landforms are produced when an event is reached on the frequency–magnitude scale of the tectono-climatic regime at which the normal behaviour is overturned and a new system created. And he continues to insist that geocatastrophes are large (with respect to the system studied) improbable events that are physically possible but which require the rare coincidence of favourable control and are therefore unlikely. He also presents the classification of geocatastrophe (Figure 2.3) and some examples. Concerning his presentation, I would like to agree in general terms but I am not willing to invoke an old term such as catastrophe (Yatsu 1992).

Recently, the expressions, catastrophe, catastrophism or neo-catastrophism have become more prevalent in geomorphology. One of the factors encouraging this trend is the development of information theory, in which catastrophe theory, e.g. cusp theory, is a language to account for smooth transition from one state to another as well as abrupt changes. Among geomorphologists, there are arguments for and against catastrophe theory.

Here, I would like to evoke the suggestions given by Robert H. Dott, Jr, in his 1982 Society of Economic Paleontologists and Mineralogists presidential address (Dott 1983).

> **Catastrophic** has become popular recently in sedimentology because of its dramatic effect, but it should be purged from our vocabulary because its use feeds the creationist–neocatastrophist cause ... Because of this creationist revival, semantics has become so important that

> **catastrophic** should not be used to characterize any geologic process — no matter how sudden or violent — that can be explained by rational scientific analysis ... It is my position that sudden and violent — as well as very slow and nonviolent — episodes are not at all exceptional geologically and are to be explained as wholly natural.

Strongly supporting his suggestion, I would like to advance a proposal that the usage of the term *catastrophe* should be limited to the ancient discussion of catastrophism and the principle of uniformity be strictly understood as the principle of nature (Yatsu 1992).

Proposition 8 analyses perturbing displacements and their results. The approach looks like a system analysis of the impulse of changes and of resulting new characteristic states. The taxonomy of geomorphic events and their relations is the main objective.

Proposition 9 explains the sensitivity of landform change and enumerates the possible barriers to geomorphological change. And it analyses thoroughly the ratio of the magnitude of the barriers to the magnitude of the disturbing forces. Accordingly we must appreciate that Proposition 9 succeeds in explaining the landscape stability very theoretically. Unfortunately it creates, together with Propositions 5 and 8, the impression that the propositions are too abstract and thus conspicuously lacking in data obtained through scientific instrumentation. It seems that Brunsden displays much interest in the subject of analysis of geomorphic systems.

Let me have a chance to present my general remarks on all ten propositions, i.e. his theoretical framework of geomorphology. The most fundamental idea is that of tectono-climatic interaction. First of all, we must be concerned with the word "tectono-". The distribution, formation and properties of continents and oceans, great mountain ranges, island arcs, orogenic zones and so on have been studied mostly by geologists so far, even though these features are also called mega-landforms. In the 1960s the concept of ocean-floor spreading and the models of plate tectonics were introduced by geophysicists, and studies of these fields have been carried out, by and large, by earth scientists who do not regard themselves as geomorphologists. Even if this model adequately accounts for many features of mega-landforms on the earth's surface, it does not appear capable of providing geomorphology with a central focus. Brunsden's idea is that the concept of tectono-climatic domain or interaction was clearly derived from the superposition of climatic geomorphology maps upon those of plate tectonics. Some geomorphologists are not only quite indifferent about the model of plate tectonics but they are also critical of climatic geomorphology. For example, Bremer (1989, p.96) explains the essence of climatic geomorphology as follows:

> Climatic and climato-genetic geomorphology were the expressions of what was done by nearly all geomorphologists in Germany in the first decades after the war. This research heavily rests on keen analysis in observation and stringent deductions in comparison and synthesis. The systematic approach asks for broad experience which causes some difficulties. Climatic as well as climatogenetic geomorphology are concerned with medium or large-scale landforms and particularly with their assemblage and interrelation requiring good observations which are hard to demonstrate and to learn. Description, mapping, and photos are the means of documentation. These are not easy to reproduce by others in other areas. Thus there is a strong subjective component.

And Baker and Twidale (1991, p.81) claim that:

> The essence of the climatic interpretation is, of course, that various climatic factors, notably precipitation, temperature and wind, induce the activity of various processes that produce assemblages of forms that are characteristic of particular conventionally defined climatic regions. Unfortunately, the various other types of climatic landform assemblages are less easily defined and identified. Some forms occur in contrasted climatic settings but are generated by common processes induced by various climatic circumstances.

and they continue their comments,

> Another criticism of climatic geomorphology is that its synthesis is premature (Stoddart, 1969). As in the case of Davis' geographical cycle, not enough is known for the details of process to assemble a rigorous scientific edifice.

Finally, Baker and Twidale declare that

> Much of modern geomorphology lacks the enchantment that the science possessed a century ago. Practical and philosophical impediments are thwarting modern attempts to achieve a satisfying understanding of landforms and their genesis.

and they propose several courses of action to revitalize geomorphology.

To my mind, climatic geomorphology is very important, but the present status of climatic geomorphology seems not to be scientific enough. I read carefully a textbook for climatic geomorphology recently published by a very famous author, and I noticed that his explanations of the climatic influences upon rock and mineral weathering are completely without experimental verification (Yatsu 1988, pp. 401–504). That is to say, climatic geomorphology today seems to be lacking in scientific precision.

Here, I mentioned "scientific", in the same sense as in Yatsu (1992). To my mind, definition, classification and scaling of the magnitude of spatial and temporal changes in landforms are not enough to make geomorphology more scientific. A great while ago, I lamented over the special customs prevailing in the geomorphological community. Many geomorphologists of that period were interested totally in questions of what, when and how (how much and how to), but not in that of why (Yatsu 1966, p.13). It seems that they still hold these most cherished traditions. The answer to the question of why is causal explanation, which is given only by science. The explanation which lacks causality is not entirely scientific.

Generally speaking, Brunsden successfully analyses the geomorphological system. He elucidates the functional and structural relationships of geomorphic phenomena, making use of the best available geomorphological literature. His work is appreciated as pedagogic material in which the valuable results of earlier studies have been incorporated. He must be an analytical philosopher. Many of those who succeed as systems analysts neither do fieldwork nor conduct experimental research. Such symptoms must be called "system analysis syndrome". It is sincerely hoped that Brunsden's paper does not proliferate this syndrome.

In short, the most serious problem of the paper is its aim and objectives. Each proposition is certainly a summary of one aspect of the present situation of geomorphology. Therefore Brunsden (1990) is one of the best chronological archives but it does not act as a motive to the progress in geomorphology. Advancement needs no mental obstacle, but creative freedom.

My wording is certainly too vehement, rude and bold. I have to apologize if I have accused him falsely, because of my serious shortage in English vocabulary. My concerns are exclusively with the future of our science and I intend no personal criticism.

CONCLUSION: A REQUEST ADDRESSED TO PRESIDENT BRUNSDEN

If a metaphor is permitted, Brunsden is like the president of some huge multinational enterprise, making the annual report to a general meeting of stockholders, on the business results including the proceeds of sales and the total stock. His report is so logical and elaborate that it is accepted unanimously and without any objection. He made a satisfactory performance of inventory control, a technique of management science and operations research for his firm. The control of inventory is of course vital to the financial strength of a firm. A careful examination of his management, however, may reveal inadequate consideration of the quality control of his products.

It was already mentioned that Baker and Twidale (1991) found the present situation of geomorphology very unsatisfactory. Earlier, Gregory (1978) advised that for physical geographers a more secure mathematical and scientific foundation is indispensable and thus amendment of curriculum for training of young students is most urgent. Considering that at present, Davisian geomorphology is still deeply rooted in geomorphological thought, I have already suggested a plan for restructuring geomorphology with an intention to invert our science (Yatsu 1992). Thus I share the same complaint as that of Baker and Twidale, and Gregory.

In any case, the premise on which President Brunsden's ideas are based still seems to be perhaps seasoned with a Davisian flavour. Indeed the name of W.M. Davis appeared 13 times in his paper. The quotation from Davis at the beginning of his paper endorses the above scepticism: *there is no smoke without fire*. The renewal of our science seems to be absolutely essential.

In the USA, during the 1960s, geomorphologists reportedly experienced something of a shakeup. Just in case it happens again, geomorphologists may need outside help. President Brunsden (1993) informed me indirectly (through Professor T. Suzuki) that Gods occasionally have association with mere mortals. If so, we may ask Brunsden for divine intervention, i.e. to build an *Ark* in order to save geomorphologists from undermining their own academic position. We may cordially invite him to issue a *White Paper* which includes an analysis and critique of contemporary geomorphology, and the perspectives and horizons of the science as he sees it into the 21st century.

President Clinton declared "We pledge that the era of deadlock and drift is over — a new season of American renewal has begun". Our own President Brunsden will hopefully announce his plan of action for the renewal of our science.

REFERENCES

Baker, V.R. and Twidale, C.R. 1991. The reenchantment of geomorphology. *Geomorphology*, **4**, 73–100.

Bremer, H. 1989. Geomorphology in Germany (The Federal Republic of). *Transactions Japanese Geomorphological Union*, **10-B**, 89–98.

Brunsden, D. 1990. Tablets of stone: toward the Ten Commandments of geomorphology. Z.

Geomorph. N.F. Suppl. Bd, **79**, 1–37.
Dott, R.H., Jr. 1983. Episodic sedimentation — How normal is average? How rare is rare? Does it matter? *Journal of Sedimentary Petrology*, **53**, 5–23.
Gregory, S. 1978. The role of physical geography in the curriculum. *Geography*, **63**, 251–264.
Yatsu, E. 1966. *Rock Control in Geomorphology*. Sozosha, Tokyo, 135 pp.
Yatsu, E. 1988. *The Nature of Weathering, an Introduction*. Sozosha, Tokyo, 624 pp.
Yatsu, E. 1992. To make geomorphology more scientific. *Transactions, Japanese Geomorphological Union*, **13-2**, 87–124.

TABLETS OF STONE: TEN COMMANDMENTS OR A GOLDEN RULE?

VICTOR R. BAKER

> And what, then, is belief? It is the demi-cadence which closes a musical phrase in the symphony of our intellectual life ... As it appeases the irritation of doubt, which is the motive for thinking, thought relaxes, and comes to rest for a moment when belief is reached. But, since belief is a rule for action, the application of which involves further doubt and further thought, at the same time that it is a stopping-place, it is also a new starting place for thought.
>
> (C.S. Peirce 1878)

SUMMARY

The advocacy of commandments or other authoritative principles for science, whether realized or as a goal, is sure to provoke strong reaction from those who view the scientific process as antidogmatic, fallible, and endlessly directed at truth. At the Second International Geomorphology Conference (Frankfurt 1989) Professor Denys Brunsden's lecture "Tablets of Stone: Toward the Ten Commandments of Geomorphology" proposed a ten-point theoretical framework for geomorphology (see Brunsden 1990). Despite its title, the framework is presented as propositions, i.e. statements put forward in a debate as to their truth or falsehood. Can such answers be provided? Will geomorphology ever have "Ten Commandments", which are presumably the right truths. Perhaps geomorphologists should be satisfied knowing "Ten Ways to be Wrong," as Professor S.A. Schumm (1991) has subtitled his book on methodology in the earth sciences. Certainly the sanctifying of any "theoretical framework" conveys the philosophical assumptions that various "cardinal" propositions are (1) possible, and (2) important for geomorphological scientific practice. The philosophical viewpoint of pragmatism is used to question these assumptions. Perhaps the Golden Rule could substitute for all Ten Commandments, though this rule's long philosophical tradition transcends even stone tablets. Of course, this is the Golden Rule for which George Bernard Shaw provided the following verbal expression: "The Golden Rule is that there are no golden rules." The Golden Rule commands diverse support from the logic of science, from pragmatic philosophy, and from common sense. Future progress in geomorphology need not lie in neo-Davisian theoretical frameworks. It is argued that this progress lies in the unimpeded desire to learn, in an endless quest served by theory, rather than commanded by it. This is the scientific spirit described by Charles S. Peirce. It is a precious spirit that requires a spirited defence.

INTRODUCTION

Central to the American philosophical tradition known as pragmatism is its critique of inquiry that purports solutions to questions of ultimate truth and falsehood. One must note that important differences occur in the philosophies espoused by the various advocates of pragmatism, which is the only modern philosophical tradition to have developed in the New World. Nevertheless, the whole movement, including familiar works by William James and John Dewey, has its modern roots in one simple maxim, formulated in the early 1870s by C.S. Peirce (Fisch 1965), but not published until 1878. Peirce (1878) wrote, "Consider what effects, which might conceivably have practical bearing, we conceive the object of our conception to have. Then, our conception of these effects is the whole of our conception of the object." Briefly, Peirce is saying that the meaning of any general idea lies not in various philosophical arguments as to its correctness, but rather in the practical consequences that flow from that idea.

A resounding corroboration of Peirce's pragmatic maxim was afforded by "Great Debate" held on 27 August 1993, at the Third International Geomorphology Conference, Hamilton, Ontario, Canada. The purpose of the debate was to discuss the paper "Tablets of Stone: toward the Ten Commandments of geomorphology" (Brunsden 1990), but the result was not particularly satisfying from the perspective of advancing the process of doing interesting science. Despite its provocative title, which had already elicited strong reaction to its religious and belief-oriented images (Yatsu 1992), Professor Brunsden's paper presented not commandments, but propositions, i.e. statements put forward as to their truth or falsehood. The paper's readers were asked whether the ten propositions could be considered to form the current theoretical framework of geomorphology, and whether any statements were moving closer to the status of commandments. If there was disagreement, then Brunsden (1990) asked the reader, "What would you write on the Tablets of Stone?"

A GOLDEN RULE

I have a short answer to the question, which I will give first, but the reasoning behind the answer will require subsequent explanation. This is because the answer is not mine, but is one that has been basic to science extending back to the ancient Greeks. What would I write on the Tablets of Stone? I would hope never to know absolute truths that would deserve such reverence, for that realization would mean an end of the quest for those truths, and that quest I know in truth to be worthwhile. Perhaps the Golden Rule could substitute for all Ten Commandments, though it hardly needs inscribing on stone tablets. Of course, I refer to that Golden Rule for which George Bernard Shaw has given the following verbal expression: "The Golden Rule is that there are no golden rules." In geomorphological form we might express this rule as follows: the one principle of geomorphology is that real landscapes are unprincipled. Let there be no mistake about this, however. The denial of principles, which are ultimate sources, origins or causes, is not the denial of continued wonderful understanding to be obtained from geomorphological inquiry. Quite the contrary! The denial is based on the conviction that such understanding can never be ultimate. The golden rule states that there are no basic, ultimate principles of geomorphology beyond which one need not seek further with inquiry. Principles can be strongly shown to occur in mathematics, and they may well occur, as claimed, in certain branches of physics, but it is not scientifically clear that they are characteristic of geomorphology. Perhaps geomorphologists should leave the *Principia* to physicists and mathematicians, many of whom have less interest in nature than in theoretical elegance. Rather than absolute, universal laws, perhaps geomorphologists should seek the natural tendencies or habits of nature. Like all habits, these most likely have a propensity for change, and the richness of that change may well be cheapened by an insistence to ascribe it to universal, timeless, ultimate expression.

It should be clear to the reader that the issue of principles being fundamental, basic, or cardinal has far less to do with scientific practice than it does with the philosophy that precedes or underlies scientific practice. Inquiry into such problems is not as common in the arts and sciences today as it was a century ago. Nevertheless, the consequences of philosophical assumptions can have major effects on geomorphological practice even though those assumptions do not receive appropriate criticism. It was for this reason that T.C. Chamberlin, whose philosophy was pragmatism (Moulton 1929), wrote that such issues, "... are basal to the earth sciences, as they are to all science, and to all true philosophy as well. The earth sciences are entitled to probe for their own bottom ... and it is altogether wholesome that they should do so" (Chamberlin 1904).

PHILOSOPHICAL ARGUMENTS

It is a general observation, corroborated by the Hamilton debate, that a large group of scientists will react to a theoretically posed position in one of two ways. Individual scientists will either agree or disagree with the stated position. However, agreement/disagreement is only one of the two possible reactions, since it assumes acceptance that the posed position addresses the right question. The second reaction is to find the original question unacceptable. Science approaches many problems in this second way. At the Hamilton debate various discussants reacted to the general question posed in Brunsden (1990) by noting the importance of certain propositions, the lack of clarity in others, the restrictiveness or incompleteness of some definitions, or the need for additional propositions. However, other discussants questioned the appropriateness of the original question, and it is this second reaction that I presented above. I have taken this position not because I feel that some alternative theoretical framework is needed, but because I sense that a strong case can be made that any theoretical framework presents the potential for a stifling influence on the creative side of scientific inquiry. Some arguments along these lines have been advanced elsewhere (Baker 1988a, 1993; Baker and Twidale 1991), and alternative positions have been presented (Rhoads and Thorn 1993). It is very important to remember, however, that these are matters of philosophy, not science. In natural sciences, like geomorphology, one can distinguish the scientific activity by recourse to some comparison against the reality of nature. Sir Karl Popper has formalized this demarcation criterion of science in his famous "principle of falsifiability" (Popper 1968) which is itself philosophical, not scientific, through self-application.

In its simple formulation, Popper's falsifiability criterion has a distinctively non-scientific component, as do all propositions argued for truth or falsehood: they restrict discussion to two-valued logic. Thus, propositions are either true or false, the activity is either science or not, and phenomena are either formative events, geocatastrophe, characteristic landforms, or they are not. However, there is a higher level of logic, more characteristic of science, known as formal argument. As matters of logic, arguments (exemplified by the Hamilton "Great Debate") rarely resolve into neat two-valued categories.

In the 1920s the Polish logician Jan Lukasiewicz worked out principles for multivalued logic in which fractional truth values are recognized between the ones and zeros of familiar two-valued logic. Multivalued logic was anticipated even earlier by C.S. Peirce (Fisch and Turquette 1966), and it now serves as the basis for the mathematics of "fuzzy logic"(Kosko and Isaka 1993). Fuzziness measures the degree to which something occurs or to which some condition exists. Complementarity is shown among the end members of a fuzzy set. That this logic is fundamental to nature is one discovery of modern physics that theoretical geomorphologists have been surprisingly slow to appreciate. The quantum world of subatomic particles, which is certainly more fundamental than the Newtonian mechanical world, achieves mathematical expression through multivalued logic.

FUNDAMENTAL CONCEPTS IN GEOMORPHOLOGY

In claiming truth or falsity, propositions have a privileged status afforded by the assumption that they are really propositions and not arguments. A demarcation principle can be applied to the status assertion, and, like much that works in the quantum world, this principle strangely affects the privileged status of the proposition. Presumably the "Cardinal Propositions" of geomorphology to be inscribed on stone tablets (Brunsden 1990) are to be considered great truths. Consider then the demarcation principle of Niels Bohr, which he considered fundamental to quantum mechanics: Trivialities are those truths where opposites are obviously absurd; great truths are those where the opposites are also great truths.

There seems broad agreement among geomorphologists that there is some special class of arguments, which can be characterized variously as "philosophical assumptions" (Twidale 1977), "paradigms" (Ollier 1981), "basic postulates" (Pitty 1982) and others listed by Brunsden (1990). Besides the obvious disagreement on what label to apply to these ideas, there is a more fundamental issue of how to express them. Baker (1986, 1988a) also listed a number of "fundamental concepts" for geomorphology. Like some, however, he called attention to the paradox that the opposites for these concepts seemed just as fundamental as the concepts themselves (Table 2.1). This raises the interesting question of whether commandments, beliefs, or fundamental theories can be discussed as propositions when their opposites are also true. By Bohr's demarcation principle, the

Table 2.1 Paradoxical fundamental assumptions in geomorphology (Baker 1986, 1988a)

Thesis	Antithesis
Gradualism	Catastrophism
Complexity	Simplicity
Morphoclimatic zonation of landscapes	Azonal landforms
Youthfulness of topography	Antiquity of relic landscapes
Randomness	Determinancy
Order	Disorder
Actualism	Past processes may lack modern counterparts
Geomorphological concern is with modern processes and modern landforms	Geomorphological concern is with the extension into the past, reconstructing ancient processes and landscapes

positions are trivial if they are indeed propositions. They are only important to the degree that they embody the multivalued logic of science.

Consider Proposition 7 in which "geocatastrophes" are held to produce new landforms by events in the tectono-climatic regime that overturn normal behaviour and create a new system. Most listings of fundamental concepts are likely to list gradualism as the norm. Catastrophes are more commonly discussed as alternatives to the prevailing view (e.g. Baker 1986). Although I have written much on the role of cataclysmic events in geomorphology (Baker 1977, 1978, 1983, 1988b), it was never from the conviction that catastrophes themselves constituted some fundamental principle. If some such principle indeed exists, it must involve a continuum of phenomena that we somewhat arbitrarily designate as gradualistic or catastrophic. The whole misunderstanding of uniformitarianism in the earth sciences (Baker 1978, 1981; Gould 1987; Shea 1982) is tied in with this question.

Ironically, given their paradoxical attributes, arbitrary classification is an unfortunate product of theoretical frameworks. In Brunsden (1990) classification is applied to landscapes (Proposition 2), to systems (Propositions 4, 5, 6) and to cataclysmic processes, or "geocatastrophes"(Proposition 7). Certainly systems are amenable to classification because of their artificiality. This is because systems are either wholly conceptual, or they are conceptually limited treatments of the perceived landscape and its processes. In contrast, real landscapes or processes have a complexity of cause–effect relationships that defies rigorous classification. Yatsu (1967) has made this point rather strongly in relation to mass-movement phenomena on hillslopes. Gilbert (1886) emphasized it in his description of scientific method, in which he describes scientific research as "... the observation of phenomena and the discovery of their relations." Gilbert (1886) talks, not of theoretical classification, but of "empiric classification", which is " ... a convenient and temporary sorting of our knowledge". Johnson (1933, 1944) develops the same theme more explicitly, and these approaches to classification continue to lie at the core of modern geological reasoning (Von Engelhardt and Zimmerman 1988).

WAYS TO BE WRONG

It is a fascinating coincidence that nearly simultaneous with the publication of "Ten Commandments of Geomorphology" (Brunsden 1990), Schumm (1991) outlined "Ten Ways to be Wrong." Schumm divides his ten arguments into three natural groupings: (1) scale and place (time, space and location), (2) cause and process (convergence, divergence, efficiency and multiplicity), and (3) system response (singularity, sensitivity and complexity). The coincidence is even more interesting in the matching of "ways to be wrong" with "commandments". Thus, sensitivity to change (Proposition 9 and a "Cardinal Proposition") and episodicity/complexity (Proposition 5 and a "Cardinal Proposition") correspond directly to Schumm's categories. Brunsden's (1990) propositions dealing with processes (3, 5, 6, 7) all encounter the difficulties of Schumm's "cause and process" grouping. Likewise, Schumm's systems response problems apply to Brunsden's (1990) cardinal propositions for resistance to change (characteristic landforms, relaxation and resistance) and propositions for tectono-climatic interaction (tectonic movements and uplift/denudation ratios). Schumm's problems of space and time apply to all supposed "principles" in geomorphology. The question is obvious. If geomorphological inquiry is somehow limited by these fundamental problems, how is it possible to achieve the perfection of a theoretical framework?

It is an interesting observation that Schumm's ten problems share many elements of various fundamental categories of logic, under which system-building philosophers like Aristotle, Kant and Hegel subsume all things intelligible. Aristotle's list from his *Organon* was also ten: time, place, position, substance, state, quantity, quality, relation, passion and action. The partial parallels to Schumm's list is interesting, but an even closer match occurs with Peirce's list of three fundamental categories, simply expressed as existence (being or mind), necessity (formal logic or matter), and representation (involving change). It is such concepts as these that have stronger philosophical justification for being a fundamental framework. Yet, even to raise this point shows how far arguments over fundamentals will drift from practical science.

METHODS OF INQUIRY

In retrospect, the term "commandment" was not particularly productive for scientific discussion. Although Brunsden (1990) did not prescribe commandments, the paper embodies the belief that propositions listed therein, or ones analogous to them, will eventually lead to "commandments," "cardinal beliefs" or a "theoretical framework" for geomorphology. Like all beliefs, this one is considered to be a basis for action, a basis of fundamental truths. Yatsu (1992) notes that geomorphology has seen such calls to action before, and he accords to them the label "neo-Davisian." But how can such a label be justified? The new call to action lacks all the old-fashioned trappings of Davisian geomorphology with its timebound, descriptive, anthropomorphic classification of landforms. Instead, all the modern systems theory–process mechanics jargon infuses each proposition. As Baker (1993) points out, such theoretical principles are commonly used in science to prescribe stipulative change in a discipline. Davis' stipulative agenda was to make geomorphology more theoretical, taking it beyond what he perceived to be philosophical weaknesses in its inductive, empirical approaches. History shows the dangers of stipulative revolutions, whose winners characterize the intellectual approaches of the losers as inferior.

Even when the allusion to God's "Ten Commandments" is removed, the suggestion of absolute authority is anathema to many scientists. In his famous essay "The fixation of belief" Peirce (1877) showed that authority is a method of inquiry that resolves doubt by an appeal to power. Less odious than authority, however, is a form of inquiry common to philosophy. This is the appeal to reason. By its infallibility, the absoluteness of its framework, principles, or beliefs, the method of reason is also incompatible with that of science. For Peirce the method of science always remains fallible as science trends, not toward some ultimate metaphysical truth, but toward a "habit of action" ultimately to be agreed upon by all who investigate. It is this habit of action, not some rule or law, that Peirce considers to be the essential quality of belief. A very similar view is argued by Bronowski (1978) in his famous book *The Common Sense of Science*.

THE FIRST RULE OF REASON

Around 1899 Charles S. Peirce wrote a manuscript entitled "The First Rule of Reason" (Hartshorne and Weiss 1931). He also noted that this rule could be regarded as the only rule of reason. He stated it as follows "... in order to learn you must desire to learn, and in so desiring not be satisfied with what you already incline to think". Of equal importance to this rule is its corollary, "... which itself deserves to be inscribed upon every wall in the city of philosophy: Do not block the way of inquiry." Peirce continues, "... to set up a philosophy which barricades the road of further advance toward truth is the one unpardonable offense in reasoning, and it is also the one to which metaphysicians have in all ages shown themselves the most addicted."

Peirce lists several ways in which "... this venomous error assails our knowledge". The first of these is absolute assertion in science. "That we can be sure of nothing in science is an ancient truth. The Academy taught it. Yet science has been infested with overconfident assertion ..."

Now it may seem that this concern with the restrictive aspect of theoretical frameworks is overblown. After all, it is not theory itself that can restrict inquiry; restriction occurs through structures in the scientific community. These are the ones that are described so well by Kuhn (1962) in his portrayal of "normal science" and "paradigms". Thus, one can argue that it is not the theory itself but rather the ideological aspects of theory that could pose some problem. Here is where we see the relevance of Peirce's pragmatic principle quoted at the beginning of this essay. In science the philosophy is irrelevant; only the consequences matter for doing science. Another way of making the point is through a well-known pragmatic sociological principle: "It does not matter if the problem is real or not. If people believe it is real, it will be real in its consequences."

CONCLUSION

There is, of course, a long tradition in science of those who refuse to ever accept truths, just as there is also a tradition of truth advocates who envision a better world by acceptance of that which is good. The latter tradition can be traced back to Plato. The lawyer-turned-geologist Charles Lyell exemplifies this second goal in his programme to change geology, making it more scientific by stipulating principles (Gould 1987). The principles, particularly substantive uniformitarianism, had restrictive consequences for understanding flood processes (Baker 1981).

Another lawyer with geological connections exemplifies those critical of ultimate truths. Clarence Darrow, defender of evolution science at the famous Skopes "Monkey Trial", always fought those excesses justified by the prevailing "truths" of his day, always inspired by his disbelieve in absolute "truths". His motto was, "The pursuit of truth will set you free, even if you never catch up with it." In this spirit, it is happy to report that in his closing academic address to the Third International Geomorphology Conference (Hamilton, 1993) Professor Brunsden discussed "Geomorphology sans frontière" by placing emphasis, not on metaphysical foundations, but on practical matters.

This whole essay has provided no new insight beyond the classic methodological writings of geologists pragmatically seeking to develop a habit of thought appropriate to their science. Mackin (1963) reiterated this "rational method" described in earlier works (Gilbert 1886; Chamberlin 1890; Johnson 1933). However, it was probably T.C. Chamberlin who best expressed the pragmatic method for the earth sciences (Chamberlin 1904): "I think we should do well to abandon all claims that we are reaching absolute truth, in the severest sense of the phrase, and content ourselves with the more modest effort to work out a system of interpretation which shall approve itself in practise under such tests as human powers can devise."

ACKNOWLEDGEMENTS

I thank D. Brunsden for the stimulation posed by his 1989 International Geomorphology Conference address, leading to my attempts at expressing the above arguments. This paper is AUMIN Contribution No. 5.

REFERENCES

Baker, V.R. 1977. Stream channel response to floods, with examples from central Texas. *Geol. Soc. Amer. Bull.*, **88**, 1057–1071.
Baker, V.R. 1978. The Spokane Flood Controversy and the Martian outflow channels. *Science*, **202**, 1249–1256.
Baker, V.R. 1981. Introduction. In: Baker, V.R. (ed.), *Catastrophic Flooding: The Origin of the Channelled Scabland.* Hutchinson Ross, Stroudsburg, Penn, pp. 1–8.
Baker, V.R. 1983. Large-scale fluvial palaeohydrology. In Gregory, K.J. (ed.), *Background to Palaeohydrology: A Perspective*, Wiley, Chichester, pp. 453–478.
Baker, V.R. 1986. Regional landforms analysis. In Short, N.M. and Blair, R.W., Jr (eds), *Geormorphology from Space: A Global Overview of Regional Landforms.* NASA Spec. Publ. 486, Washington, DC, pp. 1–26.
Baker, V.R. 1988a. Geological fluvial geomorphology. *Geol. Soc. Amer. Bull.*, **100**, 1157–1167.
Baker, V.R. 1988b. Cataclysmic processes in geomorphological systems. *Z. Geomorph. Suppl.Bd.*, **67**, 25–32.
Baker, V.R. 1993. Extraterrestrial geomorphology. Science and philosophy of Earthlike planetary landscapes. *Geomorphology*, **7**, 9–35.
Baker, V.R. and Twidale, C.R. 1991. The reenchantment of geomorphology. *Geomorphology*, **4**, 73–100.
Bronowski, J. 1978. *The Common Sense of Science.* Harvard University Press, 154 pp.
Brunsden, D. 1990. Tablets of stone: toward the Ten Commandments of geomorphology. *Z. Geomorph. Suppl.-Bd.*, **79**, 1–37.
Chamberlin, T.C. 1890. The method of multiple working hypotheses. *Science*, **15**, 92–96.
Chamberlin, T.C. 1904. The methods of the earth-sciences. *Popular Science Monthly*, **66**, 66–75.
Fisch, M.H. 1965. A chronicle of pragmatism, 1856–1879. *Monist*, **48**, 441–466.
Fisch, M.H. and Turquette, A. 1966. Peirce's triadic logic. *Transactions of the Charles S. Peirce Society*, **2**, 71–85.
Gilbert, G.K. 1886. The inculcation of scientific method by example. *American J. Sci.*, **31**, 284–299.
Gould, S.J. 1987. *Time's Arrow, Time's Cycle: Myth and Metaphor in the Discovery of Geological Time.* Harvard University Press, 222 pp.
Hartshorne, C. and Weiss, P. 1931. *The Collected Papers of Charles S. Peirce, Vol. 1.* Harvard University Press.
Johnson, D. 1933. The role of analysis in scientific investigation. *Geol. Soc. Amer. Bull.*, **44**, 461–494.
Johnson, D. 1944. Mysterious craters of the Carolina coast: A study in methods of research. *Amer. Sci.*, **32**, 1–22.
Kosko, B. and Isaka, S. 1993. Fuzzy logic . *Sci. Amer.*, **269**(1), 76.
Kuhn, T.S. 1962. *The Structure of Scientific Revolutions.* University of Chicago Press, 200 pp.
Mackin, J.H. 1963. Rational and empirical methods of investigation in geology. In Albritton, C.C. (ed.), *The Fabric of Geology.* Stanford, California Freeman, Cooper, Stanford, California, pp. 135–163.

Moulton, F.R. 1929. Thomas Chrowder Chamberlin as a philosopher. *J. Geol.*, **37**, 368–379.
Ollier, C. D. 1981. *Tectonics and Landforms*. Longman, London.
Peirce, C.S. 1877. The fixation of belief. *Popular Science Monthly*, **12**, 1–15.
Pierce, C.S. 1878. How to make our ideas clear. *Popular Science Monthly*, **12**, 286–302.
Pitty, A. 1982. *The Nature of Geomorphology*. Methuen, London, 161 pp.
Popper, K.R. 1968. *The Logic of Scientific Discovery*. Harper and Row, New York, 480 pp.
Rhoads, B.L. and Thorn, C.E. 1993. Geomorphology as science: The role of theory. *Geomorphology*, **6**, 287–307.
Schumm, S.A. 1991. *To Interpret the Earth: Ten Ways to be Wrong*. Cambridge University Press, 133 pp.
Shea, J.H. 1982. Twelve fallacies of uniformitarianism. *Geology*, **10**, 455–460.
Twidale, C.R. 1977. Fragile foundations: Some methodological problems in geomorphic research. *Rev. de Geomorph. Dyn.*, **26**, 81–95.
Von Engelhardt, W. and Zimmermann, J. 1988. *Theory of Earth Science*. Cambridge University Press, 381 pp.
Yatsu, E. 1967. Some problems on mass-movement. *Geografiska Annaler*, **49A**, 396–401.
Yatsu, E. 1992. To make geomorphology more scientific. *Trans. Japanese Geomorph. Union*, **13**, 87–124.

THE TEN COMMANDMENTS OF GEOMORPHOLOGY: A GENERAL METHODOLOGICAL APPRAISAL

JEAN L.F. TRICART

As I have a broad systems approach to landform genesis. I will first discuss that approach and then consider in turn: fluxes and landforming, the interface system, morphogenic evolution and terrestrial natural systems.

A BROAD SYSTEMS APPROACH

Physicists discover smaller and smaller particles and astronomers discover progressively more and more remote celestial bodies and galaxies. These two observations are symmetric and suggest only one philosophy. Facts which can be observed at a certain level of perception and which form a system at this level, can be subdivided into a set of subsystems which belong to a lower taxonomical level. Our level of perception is similar to the concept of resolution, as currently used in remote sensing. I do not use the word "scale" as it is inadequate and confusing. A scale is a ratio, the ratio which exists between the dimension of the real object and the dimensions of its representation on a map.

I have proposed (Tricart 1987a, 1989) an analysis of the terrestrial natural system which integrates what we know at the level of our planet. As such, geomorphology is one of its subsystems. Geomorphology is situated at the meeting point of three groups of influences:

(a) *Geological influences*, which originate inside the planet itself and, as such, are usually coined "internal forces". They can be subdivided into tectonic movements, metamorphism, and volcanism, which are interdependent. We ought to be aware that such a geological package influences geomorphogenesis at different time levels:
   (i) Immediately, mainly under the form of catastrophic events, as earthquakes or volcanic eruptions. The word "catastrophic"[1] refers to the society and is justified by both the brutality and importance of the event and by the victims and damages it causes.
   (ii) After a certain duration and during a period which can eventually be billions of years. This influence is expressed through lithology, i.e. the characteristics of the rock material in which landforms are carved. The resulting detrital material is deposited in lower areas, usually subsiding ones. The transformation of original rocks by metamorphism is obvious, just as are the specific characteristics of the material generated by volcanic eruptions. As sediments, they undergo diagenesis which modifies their initial characteristics, and, thus, influences the behaviour of the external agents of weathering, and erosion.

[1] I consider that it would be unwise to adopt the position espoused by E. Yatsu. His historical reference to the scientific use of the word "catastrophe" can be accepted only by specialists in the history of science. It is contradictory with the current use of both people and media. It is a principle of mine to use as far as possible a vocabulary which can be understood by everybody. Otherwise, we impede the social objectives of science.

(b) *External forces*, i.e. forces whose generation lies outside the planet. They integrate various phenomena which take place in the solar system, such as solar radiation, Newtonian attraction of the sun and the moon, meteoritic impacts. Meteoritic impacts ("astroblemes") were numerous and big during Precambrian times. Since then, they have been largely obliterated by an intensive morphogenic activity. This has not happened on Mars, where an early loss of the atmosphere reduced the morphogenic forces, so that an intense "craterization" remains a determinant topographic feature of the planet.

The Newtonian attraction of the moon obeys two opposing effects: from one side, it is favoured by the relatively short distance moon–earth and, from the other, reduced by the very modest mass of the satellite. The attraction of the sun results from an enormous mass outweighing the effect of a greater distance. The attraction of the moon and of the sun determines tides. The effects of the Newtonian attraction are either localized along the coasts or of restricted geomorphic importance (terrestrial tides).

Much more important are the effects of solar radiation. The web of interactions which characterizes the solar radiational system is much more varied, multiform and complex than the web of the geological forces. According to a systems approach, it can only be compared to the web of anthropogenic activities, which will be presented later. In the upper atmosphere, terrestrial magnetism deflects radioactive particles which are heavily loaded in energy: α, β, γ particles (Van Allen Belt). Nearer to the earth's surface, some 30–20 km away from it, in the stratosphere, UV radiation is absorbed by ozone (0_3), another protection for living beings. Immediately above the earth surface, the troposphere absorbs a part of the calorific radiation (thermal IR), which is the "motor" of meteorological phenomena, and therefore determines the climates on the earth. This absorption depends principally on the concentration in the atmosphere of CO_2, CH_4, H_2O vapour, aerosols and nitrogen oxides. A part of it is determined by natural mechanisms, i.e. CH_4 emission by hydromorphous environments (swamps) but another one, rapidly increasing, depends on human activities, e.g. CH_4 emitted by the cultivation of rice or by cattle. Some apocalyptic fears have been aroused about a so-called "greenhouse effect", which would result in a dramatic increase of temperature within a few decades. Presently, it is impossible to support such fears with objective data. Nevertheless, the problem deserves scientific examination, as was decided a few years ago by the International Council of Scientific Unions (Tricart and Kiewiet de Jonge 1992).

On the earth surface, the incident radiation determines temperature and potential evaporation. The infrared wavelength transmits the greater amount of calories, so that living beings have to reduce the resulting tendency to an increase of their temperature by transpiration, which absorbs calories. Nevertheless, transpiration depends on one side on the concentration of gaseous H_2O molecules in the air, and from the other on the quantity of liquid H_2O molecules available to be extracted by plant roots or drunk by animals. Water balance is therefore directly determined by climatic factors and, indirectly, by the metabolism of living beings.

The metabolism of plants, which are the unique primary producers because of their ability to synthesize cells consisting of carbohydrates using mineral matter and the energy of solar radiation in the visible wavelength (light), is typically dependent on the interface atmosphere/water or land surface. This interface can be considered as a subsystem of the terrestrial system. It is characterized by a complex interaction of life, external and internal forces. On this interface, geomorphology and life are twin phenomena.

(c) *Productive human activities are localized, too, on this interface.* Nevertheless, the wastes they produce spread themselves into the whole of the terrestrial environment. As the fluxes of the various wastes have been already presented (Tricart 1987b, 1989), I will limit myself to a partial enumeration: CO_2 emissions result from fire-clearing of agricultural land, domestic heating and cooking of food, and from industrial activities; CH_4 is produced both by agriculture and the use of fossil hydrocarbons; and CFCs and O_3 act as "greenhouse gases". These are only some of the main producers of acid rain, which are enhanced by the refining of various non-ferrous metals, among others nickel (e.g. around the nickel refinery of Sudbury, Ontario). In environments lacking materials like calcium and magnesium carbonates which can neutralize the H ions, acid rain, whose pH can be as low as 4.5, turns lake water strongly acid: pH as low as 4.2 has been monitored around Sudbury. Weathering preparation, the incipient stage of morphogenesis, being altered, the totality of morphogenetic processes and of landforms then is influenced.

Human activities also directly alter the environment through urbanization, technology and agriculture. For example, the Transalaskan Pipeline has been raised above the ground surface, at much highest cost, in order that the crude oil, which must be kept at above 0°C temperature, would not provoke the melting of permafrost. If it had not been raised, severe ground disturbances by thermokarst development would have dismantled the pipe and resulted in oil-spills, and ecosystems would have been drastically disturbed. In such an example, the combination of both ecological and geomorphological processes is evidenced.

Changes of land-use affect, in particular, water balance, local climate, soils and morphogenesis.

During and immediately after World War I, in the USA, monoculture of grains was extended in the West far beyond acceptable limits. In this semi-arid area, precipitation is highly irregular and inadequate for crops one year out of three. As long as the international situation maintained prices high, a risk remained acceptable. But around 1925, the economic situation changed. Europe's productive capacity being restored, prices fell. Simultaneously, land degradation enhanced the effects of climatic hazards. Monoculture was aimed at maximum profit, disregarding any sustainable productivity. Impoverished soils became less and less productive and the effects of drought made the vegetal cover sparser and sparser, more and more open and unprotected. The soils became more and more intensely affected by both eolian and hydric processes. It resulted, during dry spells, in drastic deflation, generating the famous "dust bowls". Fine particles deflated in the Great Plains, obliged New York residents to use electric light in the offices and the streets at noon. In the West, thousands of square kilometres of agricultural land were turned into active sand dunes. During showers, the impacting raindrops encrusted and hardened the soil surface, making it quite impermeable. Gullying, even on smooth slopes, destroyed arable land. The New Deal was the first integrated package aimed at changing agricultural practices and at soil and water rehabilitation and conservation. The cost of World War II put an end to these innovations. This example evidences the interaction of economics, technology and environment. The application of a general systems approach helps us to understand the data.

An intrinsic property of systems is that they are dynamic. They undergo perpetual changes as a result of the transformation of energy inputs through the processes which characterize their different components, i.e. subsystems. A system ought to be characterized by the fluxes of energy/matter by which it is structured. Let us now examine these fluxes, focusing on those which are of relevance for morphogenesis.

FLUXES AND LANDFORMING

Lavoisier's principle can be expressed as: "nothing is created, nothing is destroyed in Nature". For geomorphology, its application is "the substance which is transported away from a landform is to be found in another place where it contributes to building another landform". In any system, interactions occur only when a certain threshold-value of the impact is attained. This concept can be found in Brunsden (1990): some of the so-called "barriers" listed in Proposition 9 (p. 24) can be considered as thresholds. The same threshold concept can clarify the following sentence: "morphological resistance, e.g. flat slopes, low relief, rock attitude ..." Unfortunately, these components of the system are enumerated among others which are of a quite different nature: this, again, is a regrettable confusion. The application of a correct systems approach would have allowed Brunsden to escape such criticism. The same criticism can be made of the use of "rock attitude" and "rock type" (p. 24), which are both components of geological structure. "Strength resistance, e.g. rock type, weathering rate" are quite different one from the other. "Weathering rate" must be examined as an interaction in which, directly and indirectly, climate is determinant. This will be discussed later, in the next section on the "Interface System". Let me just remark here that the weathering rate depends on climate, vegetation, topographic position and water circulation either on the land surface or within surficial material. Such a rapid enumeration indicates that the problem can be examined best by using a systems approach: weathering is a subsystem of the interface system.

When following the systems approach, we must consider that a group of landforms where a material flux originates and the group of landforms where it ends together form a system, which I speak of as an association of landforms. It is defined by a flux or by various fluxes of material or debris, the word "debris" being used for any size of particles, including ions. Associations of landforms can consist of groups of landforms as above or, at a lower taxonomical level, of "individual" landforms, such as an active slope with an upper "free-face", a middle segment where material is transported downwards and a lower, terminal segment, where it is deposited as a colluvial form. Such a linkage is quite well known and accepted by geomorphologists.

In a geomorphological system, the differential interactions change according to both space and time. In space, they are different from one place to another at the same time. In time, they are different, at the same place, from one period to another. Interactions can change not only in value but even in direction, i.e. the input comes from another component, both with time and space. Every material flux is a response to an energy input. In his Proposition 3, Brunsden invokes the influence of sea-level changes. I must nevertheless confess that his paper gives me the impression of some nasty smell of the Davisian cycle of erosion, as taught by his first mentor, Professor S.W. Wooldridge. Sea-level changes are considered as a determining factor of land forming, operating universally.[1] Further on, Propositions 4 and 5 seem to be contradictory and, as such, impossible or very difficult to reconcile. In Proposition 4, "... through the operation of a constant set of processes, there will be a tendency over time to produce a set of characteristic landforms", while, according to Proposition 5 "Landforms are continually subject to perturbations which arise from changes in the environmental conditions of the system. These impulses are episodic and complex in nature at all scales. Therefore changes to landforms will be episodic and complex" (Brunsden 1990, p. 14). Perturbations are not compatible with "the operation of a constant set of processes". I do not consider that there is any automatic impulse which would, in all cases, determine landform shaping. Two different types of fluxes, progressive and regressive fluxes, must be distinguished, according to the manner in which they migrate in a catchment.

(a) *Progressive fluxes* begin in the higher parts of an area and the transported material is accumulated downstream. With time, this accumulation invades lower and lower areas. The concept can be applied either to an association of forms or to a single slope, as described above. In a very general manner, progressive fluxes result from an exportation of material which becomes gradually excessive for the capacity of transportation of the processes involved when the distance from its source increases. Even in exoreic regions sea-level changes have only a very limited influence on them, and frequently, none at all. Frequent examples are given by western European periglacial aggradation on the present sea-floor, when sea level was a hundred metres or more below the present one, for example the Würmian (Devensian) coating of the Channel floor by the alluvium of the Somme, the Seine and other rivers. Paradoxically, it has frequently happened that, while progressive accumulation took place when the sea level was low, an incision of the deposited alluvium occurred later, when the sea level was rising, during the Flandrian transgression. Catastrophic events can initiate progressive fluxes and later incision into terraces. The enormous terraces of the Venezuelan Andes are considered to result from recurrent hurricanes responsible for gigantic landslides during Quaternary cold periods. That material has been incised later during periods of modern-type climates, and is formed in a strongly upheaved geological unit. These examples focus the role of climatic changes: it is frequently "formative" in the sense of Brunsden. Effectively, the climate is of first importance for determining the size of the particles involved in the fluxes: the coarser they are (mechanical weathering being overwhelmingly predominant), the greater is the energy needed for moving them. When climatic conditions allow exportation that consists mostly of ions, they can be transported even if the slopes are very, very low. Classical examples of morphogenesis resulting from progressive fluxes are the monadnocks of interfluves, or the inselbergs cropping out of pediments.

(b) *Regressive fluxes*: I propose this name referring to the expression "regressive

[1]C.R. Twidale demonstrates brilliantly, in his own contribution, that endoreic regions extend over enormous areas so that the marine base-level cannot be considered as a planetary factor of geomorphogenesis. It is, in fact, just a special case. This is the reason for which I use "base-level", a more comprehensive notation.

erosion" widely used by Davisian geomorphologists. The term "regressive" is used because it is readily intelligible, but it is in no case a demonstration of allegiance to Davisianism ... Regressive fluxes, like regressive incisions, develop from the lower parts of the catchments towards their higher ones. Their progression is from streams of higher order to streams of lower order. A lowering of the sea level can favour regressive fluxes, but this is not always the case, as some other conditions are needed. For instance, a sufficient increase of the slope is needed. It results in an additional energy input in the system, which is used for producing the work that the stream bed incision represents. If the abandoned sea-floor is a plain sloping too gently (the inclination is lower or equal to that of the stream bed), no incision is possible as a consequence of the lack of any energy input. If a suitable material exists, an aggradation can occur, instead of an incision (cf. above). Regressive incisions, exactly as progressive fluxes, are highly dependent on the dialectical opposition between the geomorphological processes and the geological background, mainly its lithological component. This general statement is in complete agreement with Brunsden's philosophy (Brunsden 1990).

The above statement is focused on the spatial aspect of fluxes and, as such, needs to be complemented with a temporal aspect. It will be limited now to short durations as longer ones will be examined under a later heading, devoted to the evolution of landforms. A characteristic of all material fluxes is their discontinuity, in space and time, which are intimately interdependent. The word "colluvium" has been adopted by geomorphologists for designating fine material deposited at the foot of a slope from which it originated. It is nevertheless also an accumulation characterized by the two following aspects:

(i) A geologically short duration, spanning from 10 to a few 10^3 years. It does not eliminate deposits which are much older than this, as colluvia have been formed during any geological period, but this time span refers to the duration involved in their formation.
(ii) A sufficiently fine material: what the soil scientists define as "fine earth", i.e. a mixture of particles < 2 cm, containing only a few larger particles. It would be quite improper to include stone and boulders under the name "colluvium".

These two restrictive conditions ought to be respected together for a proper definition of colluvia. The reference to granulometry is especially important for geomorphology, as it involves the concept of competency which designates the upper threshold for size and weight of the aptitude of a fluid (air, water, plastic material as earth and ice) to move particles. The laws of similarity necessitate the use of completely artificial material in flumes for hydraulic experiments, so that competency has not been thoroughly investigated by hydraulic scientists. It is nevertheless of prime importance for geomorphology. Effectively, it can be evaluated with acceptable accuracy from field measurements, using a quite simple approach invented more than 30 years ago (Tricart and Cailleux 1953). In fact, it is an extreme manifestation of the differential speed of the particles in fluxes which determine what can be named "mobility" (Tricart 1961, 1962). Unfortunately, this decisive problem has been jumped over by Brunsden when he writes, "in these models denudation is a function of elevation. No allowance has been made for minor fluctuations in net rates over time due to a climatic change, etc. The models are thus very general" (Brunsden 1990, p. 12). Landforming disregarding any flux of material is trying to make chemistry experiments without any chemical product. Such "models" are not "very general", but useless. Moreover, it is in complete contradiction with Proposition 5 of Brunsden: "Landforms are continually subject to perturbations which arise from changes in the environmental conditions of the system ..."[1]

Differential particle mobility depends not only on the characteristics of the particles: it depends too on purely geomorphological conditions, i.e. the site where they have been deposited in transit. For example, colluvia can be soon mobilized again when laid near a river bed where banks are frequently sapped. They will be reworked after a much longer timespan if they can be reached only by exceptional floods. Particle fluxes are a kind of relay race, but with no simple and permanent rules. Remobilization of particles depends on so many intricate rules which interact one with the other that any construction of theoretical models is out of reach, because they can only be oversimplified and thus inadequate. We must practice observation to be able to make correct measurements of the interface system.

THE INTERFACE SYSTEM

The interface system is defined as the main area characterized by the opposition of internal and external forces. The interface atmosphere/lithosphere, where landforms are shaped, is also characterized by the development of life, which results in a chain of effects: soil formation, rock weathering and influence on water balance.

Vegetation influences the climate by the following processes:

(i) By producing "rugosity", which reduces wind speed and results in a dispersion of a part of the wind energy in the form of turbulent eddies. It can cause a stronger transpiration of the vegetation, which can partly at least compensate it by an increased extraction from the soil, so that it has repercussions on the soil water balance. These eddies are also able to mobilize dust and eventually to scour out loam and sand from the places devoid of vegetation.

(ii) By affecting the energy of the solar radiative influx: the thermal radiation is kept, as far as possible, under control in order that the temperature of the plants should remain below the lethal threshold. This control is achieved through transpiration, which uses a part of the incident radiation to transform liquid H_2O into water vapour. Around the plants, the air is somewhat fresher and contains more humidity. The radiation in the visible wavelength is partially absorbed to be used for photosynthesis, which maintains the phytomass. As a counterpart, decaying organs of the plants fall on the ground surface where they are the condition of the formation of soils, which play an important role in geomorphogenesis.

(iii) Interception of precipitation is another type of influence of vegetation on the climate at ground level. It affects both rain and snowfalls, although it differs somewhat in the two cases. Snow usually melts slower in woods than in open terrain. In a

[1]Some fierce criticisms of D. Brunsden's reasoning can be found in the contributions of my colleagues, which I have been able to read only after the Round Table, as a courtesy of O. Slaymaker, to whom I am deeply indebted.

catchment which is partially forested, spring high waters last a longer time, but at the cost of lower discharge peaks. The magnitude/frequency curve of discharges is modified and, as a consequence, so are the river dynamics and their landforming activity. On another side, interception water is subtracted from the "climatic" precipitation and does not participate either in pedogenesis or rock weathering and karst development.

The most important component of the interface system consists of the climate–landform relationship. A distinction must be introduced between two spatial scales as follows:

(a) *Macroscale units — Topography and orography.* These are the big topographic and orographic units such as the alignment of marine troughs along the contact of the overriding South American plates and the plunging Pacific plates, which determine the disposition of the relief units. Antarctic cold waters reach the Galapagos Islands, under the Equator, and the Andes mountains form a nearly unpassable barrier for lower air masses. This structural influence almost completely overwhelms any climatic zonality. The coastal plain of the Pacific is arid and subarid, with relatively low temperatures which result from important upwellings. Weathering and other morphogenic processes are very special, so that specific landforms associations are developed, just as endemic plants were born (e.g. *Tillandsia*, which does not have any roots). Contrasting with this coastal misty desert, on the eastern flank of the Andes and on their piedmont are superhumid climates where precipitation is over 5000 mm/year in various places. The mountainous chain forms a very efficient barrier opposed to the circulation of air masses as its altitude is everywhere above 3000 m. Zonality is nearly completely obliterated from the Equator to south of the Tropic of Capricorn. The Alps, the Himalaya and the Rockies are other examples of the same type of influence of orography on climate, even if they are less extreme than the South American Andes.
(b) *Mesoscale units — Local climates.* At a much smaller dimension, that of hectometric or kilometric landforms, there cannot be different regional climates. Climatic networks have a much too low density to be of any use. In these conditions, the investigators themselves must equip plots in the immediate vicinity of instruments to densities that they consider adequate. There is no standardization either for the measurements themselves or for their tabulation, presentation and publication. Generalization is difficult and the data are frequently difficult to obtain. Nevertheless, climatic parameters which are absent from the official climatic yearbooks, such as intensity of showers, temperature on the surface of a bare soil or below the vegetation and at different depths are usually measured. Such data are specially devised for investigations of processes, and are thus very useful for geomorphology. Aspect is the main factor, as it determines the quantity of energy that solar radiation yields to the geomorphogenic system. In the Alps, the peasants traditionally oppose the "adret" or "sonnenseite" to the "ubac" or "schattenseite", i.e. the slope looking towards the sun to the one which remains in the shadow. At such middle latitudes, on the sunny slopes the altitudinal limits of the plants are higher and the growing season longer. Obviously, geomorphic differences are associated with such an agricultural contrast. In particular, soil freezing, snow-melt and water regime differ according to aspect and it is exactly the same for processes such as frost heaving, geliturbation,

gelifluction, solifluction, rain splash and runoff. Soil conservation, if required, necessitates different solutions because it is dependent, on the one hand, on the natural environment and, on the other, on agricultural systems. The thermal regime is not the only consequence of aspect, even if it is the most important. Some slopes are in a windward position and the opposite ones in a leeward position. When rainfall is associated with strong winds, precipitation can be significantly different on both slopes. Perhaps more important are the differences occurring during snowfalls. Near the divides, mostly when they are gently convex, snow cannot accumulate easily on the windward slope. Most of it is drifted towards the leeslope where it accumulates. The geomorphogenic consequences are a snow oversupply on the leeslope able to cause snow avalanches, which enhance the possibility of glacier formation. On the windward slope, on the contrary, the snow mantle is discontinuous, so that periglacial processes are the leading processes of the geomorphogenic system. Snowmelt water cooperates with them, as the small patches of snow which have been formed in the places where wind is weaker melt as soon as the cold season is over. Such a contrast is the rule near the climatic limit of glaciers. It can be observed at present in mountains and at high latitudes, but its location was lower, altitudinally and latitudinally, during the Quaternary cold periods. Ice characteristics and dynamics are under the combined influences of wind and thermal regime, both of them depending on aspect at all latitudes.

One of my former principles is heavily supported by these observations: the importance of interrelations changes according to the dimensions of the units considered, and even the direction of vectorial causalities can be reversed. At the perception level of hundreds of kilometres, the relief, the orography, determines climates; at the hectometric level, on the contrary, aspect modifies the morphogenic system which shapes the landforms. This principle contradicts Proposition 7 "New landforms are produced when an event is reached on the frequency–magnitude scale of the tectono-climatic regime, at which the normal behaviour is overturned and a new system is created. Such events are called geocatastrophes" (Brunsden 1990, p. 21). Changes of geomorphogenic systems do not require "catastrophes": they may occur under steady-state regimes, but under defined climatic conditions and in specific geomorphic sites. This conclusion introduces the last point I intend to speak about:

MORPHOGENIC EVOLUTION: GEOMORPHOLOGY AND TIME

Before any discussion of morphogenic evolution, a basic point ought to be examined. What are the durations needed for the formation of the various components of the terrestrial natural system; how long are they able to maintain themselves with their specific features; and how long is the time needed to obliterate them completely? Unfortunately, this problem has been ignored in the literature and I am myself at the beginning of my reflections. Brunsden refers, in his Proposition 7, to some statements of Ruxton (1968), which pretend to determine approximate durations of catastrophic events (see Figure 2.3). I consider them to be quite unacceptable, first because, by definition, a catastrophic event is of a very short human duration: a flood, an earthquake, a hurricane do not last more than a few days or weeks; second because the list given is an irrational succession of phenomena without any relation one with the other. For example, an orogeny has not the least similarity with a "climatic change" or a "river channel change". This last one just may be, at best, a result of the former, but can be caused also by many other factors which are not listed. Giving a duration of 10^8 years for ice ages and of only 10^7 to 10^8 years for orogenesis is worse than irrational. Orogenesis needs more time than a glaciation and there are many glaciations, more than the "classical" four alpine ones of Penck and Bruckner, during the Pleistocene, which began 2.4 to 2.3 millions years ago. All this was already known in 1968, when Ruxton published his paper. To recopy this is even more irrelevant than it was 25 years ago.

As Brunsden could not refer to my own reflections about this fundamental problem, as they have been published after he wrote his own "Tablets of Stone", I will discuss his position before giving a summary of my reflections.

We share together, D. Brunsden and I, a major concern, which is situated at the very centre of geomorphological philosophy. How and when do changes occur in geomorphogenesis, i.e. landform evolution? Brunsden's answer to these questions has been strongly influenced by the Davisian position of Wooldridge. It requires efforts that Brunsden permanently makes to struggle against this cyclic explanation. But I find that the success of such struggle has remained only partial, so that I have to criticize some of his positions.

(a) *Proposition 4.* If I do agree when Brunsden writes "... a set of forces whose variation through time can be described by magnitude and frequency distributions", I am obliged to oppose the statement that they can remain constant through time and lead to the production of an "uniform repetitive assemblage of landforms" as well as the "... strong interdependence of process and form at all scales". Exposing the systems approach and the dynamics at the interface, I have already argued against what I consider as unacceptable in these propositions. The same morphogenic system does not compulsorily produce repetitive assemblages of landforms, but associations of landforms which are diverse and changing through time and space. The interdependence of process is not uniform at all "scales", for it does change according to the very dimension of the unit considered. The excessive "uniformitarianism" adopted by Brunsden is a monadnock left by the sticky cycles ... he cannot get rid of. This cyclic prejudice manifests itself very clearly in the scheme of Brunsden's fig. 2. The initial slope is very steep, indicating a rapid change, just as the tectonic deformation provoking for Davis the beginning of a new erosion cycle was quasi-instantaneous.

(b) *Propositions 5, 6 and 8.* Proposition 5 "available data indicate that over timescales up to c. 10^5 years stability of landform controls is rare and seldom lasts more than 5–10,000 years ... only the most mobile systems (soft rocks, steep slopes, fluid beds, etc.) can normally reach a characteristic state with contemporary conditions" (Brunsden 1990, p. 14). For me, as already expressed, the problem consists of the ratio between the energy input in any system and the energy required to provoke a change in it, i.e. a qualitative/quantitative modification of the fluxes structuring this system. No generalization is possible and the one attempted by Brunsden is excessive. The factors invoked belong to various different taxonomy levels and must not be put together. The abrupt dichotomy which is fundamental in Davisianism appears again when Brunsden invokes the obsolete position of Erhart which is to oppose "biostasis" and "rhexistasis". Again, "long periods of boredom interrupted by brief moments of terror" (Brunsden 1990, p. 17) is pure Davisianism, even in the vocabulary. The same Davisian mentality guides Proposition 8: "When a perturbing displacement exceeds the resistance of the system, the system will react and relax towards a new stable state which will be expressed by a new characteristic form" (Brunsden 1990, p. 22). The first part of this proposition conforms with the Davisian axiom according to which, when a cycle is interrupted, a new one begins. Nevertheless, the second part differs completely from the Davisian cycle concept. The evolution, after the interruption, instead of again taking the road towards peneplanation, tends to elaborate another type of landforms. I could summarize briefly the comparison of the positions of Davis and Brunsden in the following sentence: abrupt beginning of the cycle: yes! finalism in the evolution of landforms: no! But the divorce is not clearly accepted by Brunsden, who immediately looks behind him: "Major change or geocatastrophe[1] will occur if the induced change crosses a critical structural specification of the system" (Brunsden 1990, p. 22). I cannot agree more with this affirmation, "Of course, the non-formative events are not really redundant because they do some work, alter the susceptibility of the system and may thereby lower the frequency and magnitude of the event that actually causes a formative change ... the main point, however, is that these small events do not appear to have an immediate or direct effect on form" (Brunsden 1990, p. 20). How can phenomena be understood which do certain work and do not affect the landform on which this very work is done? Process without fluxes? I do not understand at all. Proposition 6 contradicts the statements quoted above: "Each geomorphological tectono-climatic regime is characterized by a nested hierarchy of such effective 'process' or 'control' events distributed in time and space and described by frequency and magnitude distributions. The series of events for each regime is unique" (Brunsden 1990, p. 18). Finalism is clearly abandoned in favour of diversity. The complexity of the terrestrial natural system has already forced me to consider diversity to be a fundamental characteristic of geomorphic systems. Nevertheless this statement contradicts that which is cited above, i.e. "uniform repetitive assemblage of landforms". Catastrophism reappears in the Tablets of Stone, catastrophism is the departure point of the Davisian cycle of erosion.

(c) *Proposition 7*. "More rarely a pulsed event may occur on a massive scale. These events are believed to be capable of completely changing the system and are therefore called catastrophes" ... "landform change takes place when periods of slow change and characteristic form (stasis, stability, steady state, equilibrium) are upset by sudden discontinuous formative events" (Brunsden 1990, p.20). "If the term 'event' is thought of as being a period of increased geomorphological activity during which landscape change takes place, then the hierarchy of "events" can be constructed over all time and space scales. This might range from an orogeny to a high-magnitude storm. The hierarchy might then be subdivided into cycles, epicycles, phases, episodes, events and sequences depending on the nature of the activity" (Brunsden 1990, p. 19). "Pulsed

[1]Such an abuse of the term "geocatastrophes" resulted in the exasperation of E. Yatsu and his consequent rejection of the word "catastrophe" itself!

changes are short, extreme, episodic (on all time scales) events which cause sudden, massive energy pulses and work loads. They are followed by a return to, or near to, the initial stage" (Brunsden 1990, p. 19). These citations demonstrate the Davisian concepts which persist.

(d) *Proposition 10.* To finish with, there is an intense smell of Davisian peneplains in Proposition 10: "the ability of a landscape to resist impulses of change tends to increase with time". Davis' cycle begins with a sudden, timeless change (catastrophe) and declines asymptotically into a peneplain. The equivocal position of Brunsden enables him to gain the support of the living fossils of Davisianism as well as progressive thinkers. The new elements and perhaps, even more, the new terminology used, can give him a progressive stature. Time remains too short to fulfil a Darwinian extrication.

TERRESTRIAL NATURAL SYSTEMS

Table 2.2, which summarizes the time scales of the components of terrestrial natural systems, requires the following considerations:

(a) Human impact can be a very efficient accelerating factor. It could be said that the main superiority of the pretentiously named *Homo faber* over apes is that man has invented the saw, that apes ignore, and thus he is able to cut off the branch on which he is sitting when apes are unable to do so.[1] One of the more visible effects of human impact in the landscape is the destruction of soils by poor agronomical practices, such as deficient drainage of irrigated fields, which result in hydromorphism within 10–30 years, or in alkalinization and salinization, which enhance deflation and eolian fluxes

TABLE 2.2 Time scales of the components of the terrestrial natural system

Earth system component	Order of magnitude (years) of time involved in their		
	Formation	Persistence	Destruction
Orogene (folded mountain chain)	some 10^7	a few 10^7	some 10^6
Fault scarp, some 100 km long	10 to 10^4	10^4 to 10^5	10^3 to 10^4
Living species	$\geq 10^6$	$\geq 10^7$	10^5 or 10^1–10^2*
Biocenosis	10^4 to 10^5**	$\leq 10^4$**	10^1* to 10^3**
Weathering products and soils	10^1 to 10^6	10^4 to 10^7***	10^1 to 10^4–10^5
Frost splitting	10^{-2} to 10^2	10^3 to 10^4	10^{-2} to 10^2
Travertine dam	10^1 to 10^4	10^4 to 10^5	10^3 to 10^4
River terrace	10^4	10^4 to 10^6	10^3 to 10^4
Decametric gully	10 to 10^2	10^3 to 10^4	10^2 o 10^3
Calcrete	10^3 to 10^4	10^5 to 10^6	$\geq 10^4$
Ferricrusts ("laterite")	10^4 to 10^5	$\geq 10^6$	$\geq 10^4$

* when resulting from human impact
** when resulting from climatic change
*** if both mechanically and chemically resistant

[1]Human impacts on nature have been at the core of the Man and Biosphere Program of UNESCO and again are the core of the International Geosphere.

of particles, or such as tillage of excessive slopes, which produces massively increased detrital fluxes. Degradation of soils is even more dramatic when confronted with underdevelopment and the present demographic burst.

(b) During the Pleistocene, the geological history of the earth has been characterized by climate changes of an exceptional amplitude and frequency when compared to former times. The same type of fluctuations may have occurred at the end of the Proterozoic and during some periods of the Paleozoic, but we lack any sufficiently accurate chronometer to investigate them. Such changes could result from a chaotic (in the mathematical sense) type of evolution taking the place of a former steady-state evolution. If it were such, it would preclude any precise projection for the future. That is why the ICSU's IGBP consists, during the first phase, of a minute reconstruction of the changes which have occurred on earth since the last-but-one cold period, and of the mechanisms involved. A second step would consist of the elaboration of recommendations for the political leaders. Unfortunately, the UNCED, Rio de Janeiro, cannot give us much optimism. Pleistocene climatic changes have resulted in some species' extinctions, but in a much lesser quantity than man's impact have cause during the last few centuries and this rhythm is exponentially accelerating. Extinctions are much more rapid than speciation. Before the huge increase of human impact, migrations were more important than extinctions in the planetary biogeography. At present, the situation is reversed. In spite of the development of the biotechnologies, the new varieties or species produced are less numerous than the species which become extinct or are no more cultivated or bred. The influence of such changes on the environment, in particular on water balance and detrital fluxes, have not been sufficiently investigated. My opinion is that such investigations, conducted using a systems approach, are a part of geomorphology for the future.

(c) The orders of magnitude of Table 2.2 differ significantly in the three columns. With a few exceptions, the time needed for the "formation" of a component of the terrestrial natural system is longer than the time during which its destruction is performed. We can define this as a negative balance. In contrast, when time needed for the destruction of a component exceeds the time of its formation, the balance is positive. Unfortunately, positive balances are mostly those of components of reduced use for mankind, such as travertine dams or calcrete. More unfortunate is the case of degradational landforms such as gullies. Nevertheless, increased fluxes of particles can be favourable when they eliminate unfavourable inherited material, as in the central coastal part of Bahia State, Brazil: an old regolith there had been intensely leached and became infertile. During the last cold period, under drier climates, the phytostability was reduced, so that this regolith has been exported into the valley bottoms and the neighbouring sea. The alteration front on the Precambrian metamorphic rocks has been brought near the land surface as a consequence and the decomposition of fresh minerals yields nutrients to the roots (Tricart and Silva 1958, 1960).

Once more, I am obliged to insist on the absolute need to use a systems approach and instead of splitting the data into various commandments, to put into evidence the interrelationships that exist between them. Most of these are indirect, so that we must cross the traditional limits of disciplines, which are true blinkers, and adopt a trans-disciplinary approach. It is the only way in which geomorphology will be able to yield an effective and useful contribution to the IGBP of the ICSU.

REFERENCES

Brunsden, D. 1990. Tablets of stone: toward the Ten Commandments of Geomorphology. *Z. Geomorph. Suppl.-Bd.*, **79**, 2–37.

Penck, A. and Bruckner, E. 1901–09. *Die Alps im Eiszeitalters*, 3 vols. Touchritz, Leipzig.

Ruxton, B.P. 1968. Order and disorder in land. In Stewart, A.G. (ed.), *Land Evaluation*, Macmillan, Melbourne, pp. 29–39.

Tricart, J. 1953. Résultants préliminaires d'expériences sur la désagregation des roches sédimentaires par le gel. *C.R. Ac. Sciences Paris, 1de sem*, 1296–1298.

Tricart, J. 1961. Observations sur le charriage des materiaux grossiers par les cours d'eau *Rev. Géomorphologie Dynamique*, **XII**, 3–15.

Tricart, J. 1962. Les discontinuités dans les phénomenes d'érosion. *Publ. A.I.H.S.*, **59**, 35–46.

Tricart, J. 1985. Dialectique de le géomorphologie climatique: quelques réflexions suscitées par un "avocat du diable" bien naïf. *Finisterra (Lisboa)*, **XX**(40), 267–273.

Tricart, J. 1987a. Le Système Naturel Terrestre. *Rev. Géom. Dynamique*, **XXXVI**(1), 3–16.

Tricart, J. 1987b. Approche systémique de la degradation des terres. *Rev. Geom. Dynamique*, **XXXVI**(4), 122–127.

Tricart, J. 1989. Earth Natural System, A contribution to the Geosphere–Biosphere Program. *Geoökodynamik*, **10**(2/3), 159–176.

Tricart, J. and Cailleux, A. 1953. Détermination du centile en granulométrie. *Bull. Soc. Géol. de France, 6º Série*, **III**, 747–759.

Tricart, J. and Kiewiet de Jonge, C. 1992. *Ecogeography and Rural Management: A Methodological Contribution to the International Geosphere–Biosphere Program*. Longman Scientific and Technical, Harlow, 267 pp.

Tricart, J. and Silva, T. Cardosa da 1958. Une oscillation glaciaire carbonifère a Sorocaba (Etat de S. Paulo, Brésil). *C.R. Somm. Soc. Géol. de France*, 259–262.

Tricart, J. and Silva, T. Cardosa da 1960. Aspects de la sédimentation glaciaire éocambrienne dans l'Etat de Bahia (Brésil). *C.R. Somm. Soc. Géol. de France*, 140–141.

GEO-APOLOGIA

DENYS BRUNSDEN

INTRODUCTORY AND WITH MAIN REFERENCE TO VICTOR BAKER, C.R. TWIDALE AND EIJU YATSU

It is astonishing that the title of a paper can provoke so much rebuttal. The title of the Plenary lecture[1] at the Second International Geomorphology Conference in Frankfurt was "Tablets of Stone: Toward the Ten Commandments of Geomorphology" (The Ten C's).

Two reasons for the title were given in the Acknowledgements. First the paper is a condensation of ten lectures prepared for a course at the University of Canterbury in New Zealand in which each lecture discussed one central geomorphological idea and I was seeking an eye-catching title for the "big" lecture. Secondly, W.M. Davis' (1922) biographical memoir of G.K. Gilbert summarized the "spirit" of the paper with:

> Would that the pencilled outlines in the little pocket diaries have been written out elsewhere more at length; and yet how short would have been their endurance as the centuries roll by, even had they been engraved on tablets of stone with an iron quill.

What Davis was asking for was that Gilbert should have written out more fully the ideas he had for the development of landforms and the operation of geomorphological systems so that we could all benefit, understand more fully, debate, question and *move on*.

My wife suggested the obvious title because she understood that the purpose of the lecture was to write out our current ideas and to judge whether we had made any progress on what Baker and Twidale (1991) have called the endless quest for truth. I liked it because I find it helpful to "take stock". It is sensible to ask what we know or think we know and understand and above all to be aware of the framework of thought that we use when we study a piece of country.

Unless we are aware of the colour of the spectacles we wear when we view a scene (Chorley and Haggett 1965), how can we be aware of the degree to which our understanding is a value judgement conditioned by unexamined or unstated beliefs? It is not possible to understand a piece of research unless the conceptual framework within which the work is conceived and executed is known and yet I suspect that most of us, if challenged, would not immediately be able to state our "position".

PRINCIPLES

I agree with Baker, Twidale and Yatsu that in order to advance we need creative thought and not mental obstacles. With Twidale I feel that theoretical frameworks can form a useful synthesis, serve a strong pedagogical function, and stimulate thinking or extensions of contemporary thought as long as the system is not rigid and restrictive. For this reason I was very careful not to use the word "*principle*". The word chosen was

[1]Despite statements to the contrary the lecture was not a Presidential Address, since the IAG did not exist at that time.

"*proposition*" because it means a word put forward in debate which can be examined as to how far it is right or wrong, as far as we are able to judge with the knowledge available to us. (This is especially important when we examine the consequences of the proposition.)

The propositions were very carefully worded so that they could be falsified and rejected. This was done because G.K. Gilbert once said that "the richest man is a man rich in rejected hypotheses". It was also a response to a criticism (Haines-Young and Petch 1986) of an earlier set of propositions (Brunsden and Thornes 1979) that they were not "presented in a form which is in any way testable" and that they were insulated from any kind of refutation.

A careful reading of the Ten C's will also show that I highlighted the second part of the biographical quote about Gilbert (Davis 1922), namely that even had the ideas been etched on stone with an iron pen, the surface of a rock mass weathers and decays and the detail on it becomes blurred or even effaced. Even if the stone were erected as an edifice (or tombstone), as for example the "Cycle" of Davis and, because it was successful, remained standing for a long time, it would eventually lean and fall. It is clear that I was not seeking immortality for the ideas despite the cryptic comments of Yatsu.

It is true, however, that the use of the word "commandment" was a gross error. "Authority" and "prohibition" are not geomorphological concepts and certainly should not be applied to ideas developed in New Zealand! Certainly it was not a wish to dictate to the profession. I entirely agree with Eiju Yatsu who points out that "Commandments seem to be beyond falsification. Therefore the term 'commandment' should not be replaced by the term "proposition'."

It would have been better to call the paper "Piles of Stone". I could then have used the views of Karl Popper (1972) who wrote:

> The empirical basis of objective science has thus nothing absolute about it. Science does not rest upon solid bedrock. The bold structure of its scientific theories rises, as it were, above a swamp. It is a building erected on piles. The piles are driven down from above into the swamp, but not down to any natural or "given" base; and if we stop driving the piles deeper, it is not because we have reached firm ground We simply stop when we are satisfied that the piles are firm enough to carry the structure, at least for the time being.

The purpose was to consider which propositions form our current framework and to ask whether we have retained or confirmed any of our earlier statements. By that I wanted to discover whether we had any new ideas and whether some had stood the test of time so that we might be a little surer of them. In terms of the Gilbert quote it might mean that the etching was a bit deeper, that the rock was a bit more resistant. It might also mean that the intellectual climate and attrition processes which challenge the statements are weak? Hopefully, however, it may be that the piles are firm and that we can realistically use them to hold more advanced conversations with earth ...?

Unfortunately the use of the word "Commandment" has suggested that I wish to erect a very mandatory and restrictive structure and that I really mean my "Propositions" to be "Principles" — meaning "truth". This view has been reinforced by my use of the term "cardinal proposition" which Baker calls "great truth". In truth the purpose was to describe a grouping of propositions into more extensive ideas, such as complex cause, complex response, sensitivity to change, tectono-climatic interaction in time and space, constant process-characteristic form and persistence.

It is reasonable, therefore, for the debators to complain that some of the terminology used has been unhelpful. Despite this I defend the approach of stating my beliefs and thought processes.

Baker would prefer that, rather than seek principles, we should seek the "natural tendencies or habits of nature". This seems perfectly reasonable. For example, the proposition that processes adjust to structure is a statement of a natural tendency and it is a habit for landforms to respond to formative events. Indeed the words suit all of my propositions. Perhaps the new title should be extended to "Piles of Stone: towards the natural tendencies (or habits of thought) of geomorphology".

It will be noticed that Baker also appeals for us to "develop a habit of thought", "a rational method", a "system of interpretation", a "habit of action". Frankly, I see no difference between this and the rationale for the Ten C's other than pedantry. Actually I was suggesting that we should not just develop a system of interpretations but to recognize which one we had in case it influenced our judgement. How can we develop our habit of thought if we do not know what our current bad habits are?

My stance in this debate is, therefore, that I am sorry that my title and keywords have misled the "house" into believing that I was trying to be prescriptive. I do not understand Baker's view that we must go into the field for conversations with the earth without being influenced by pre-existing ideas. Surely Bremer must be right when she debates that to work without reference to what has gone before or without a framework of understanding is to merely end up reinventing the wheel. It is better to agree what we have already discovered and move on. I have met students who hold enchanted conversations with the earth "unencumbered by the facts" but I cannot recommend the method.

SOME SPECIFIC POINTS

H. Bremer

Hanna Bremer was asked to debate Propositions 1, 2 and 3 of the Ten C's. She has been most generous in her acceptance of the basic ideas incorporated in these propositions. Her general understanding of the aims of the proposals is also profound and there is a valuable expansion and suggestion for additional guiding concepts.

Bremer has quickly grasped the point that a system of understanding like the "Ten Commandments" is needed if we are to "bridge the gap between process geomorphology and climatic geomorphology" or indeed any set of "geomorphologies" which choose different scales of operation. This is vital if we are to successfully face "the challenge of extrapolating the short term record of measurable processes to the relatively unknown time span of, say, 100–10,000 years or beyond" (Brunsden and Thornes 1979) and to the reality of long-term landform change. Bremer recognizes that climatic, tectonic and process scales are encompassed in the scheme and are relevant at all time and space scales.

Secondly, there is a strong statement that the concepts of dynamic equilibrium, steady state and steady evolution are "assumptions which are almost impossible to prove in nature". They may be useful as a guide to relief elements of recent origin but they are of much less use or validity as we face up to the challenge of change through time. It is the nature and the controls of these changes that form the substance of the Ten C's.

Bremer also valuably asserts that the preconceived ideas that we so readily take from the systems and models of other sciences are becoming less important as we follow our own "quest for truth".

Her quest, and one of the most important contributions to geomorphology, has been the explanation (summarized in Bremer 1989) of the influence of relict landforms on present processes and landform development. The ideas are given in her discussions of divergent weathering and erosion and the idea of the traditionally continued development of landscape even after a profound change in the controlling factors, particularly climate. This work suggests that the way in which perturbations of a system are propogated is influenced by the inherited form.

We are also required to determine the relief generations, which actually can or will take place, because not all impulses of change will actually overcome the friction or inertia of the system. Bremer points out that some inherited forms become massive thresholds in the new system. Relaxation times for the process of adjustment are correspondingly very long, even to million of years.

This perhaps allows us to formulate two new propositions for discussion:

Proposition A: Landforms owe part of their character and development to the relict forms inherited from the processes of change determined by the previous tectono-climatic regimes or relief generations.

Proposition B: Landforms will continue to develop in the traditional (previous) way if the inherited form provides major thresholds and barriers to change and long relaxation times. There will therefore be a persistence of form through time.

It is perhaps worth reminding ourselves that this is a statement made very clearly by Budel (1982):

> All present processes in all climatic zones of the world act upon a stage set by ancient processes. The current processes may in some cases have behaved in the same or similar fashion for long periods of time, and may continue the activities of the ancient mechanisms with little or no change (as in the seasonal tropics). They may be traditionally shaping the inherited relief, rapidly destroying it, or having no effect on it at all

C.R. Twidale

Rol Twidale also draws our attention to the existence of ancient epigene surfaces and paleoforms. These range in age, in cratonic areas, from Archean to Pleistocene. Some predate the rifting and drifting of Gondwanaland elements and have survived since the late Jurassic. As Twidale emphasizes, such ancient features should have been obliterated long ago if we were to follow the dictates of some "well known models of landscape evolution".

It is true that the Ten C's did not emphasize the effect of time but discussed instead barriers to change, slowly responding landforms, insensitivity to change, persistence of relief, stagnancy of development and palimpsest of forms (Proposition 9). It may also be worth repeating that the last words of the paper were that "The ability of a landscape to resist impulses of change tends to increase with time. It becomes set in its ways — cast in stone".

It must be admitted, however, that the subject was not well covered and I felt compelled to expand my ideas in the "Persistence of landforms" written for Hanna Bremer's Festschrift (Brunsden 1993). Here, however, a different viewpoint to that advocated by Twidale was adopted by drawing attention to the idea of the lifetime of landforms, the length of time over which they actually persist (live) rather than their age (McSaveney and Griffiths, unpublished 1988; Griffiths 1993). Perhaps we should develop Proposition B by extracting from Budel's quote:

Proposition C: All present processes, in all climatic zones of the world act upon a stage set by ancient processes.

Twidale brings us back to a full consideration of the processes which act through time. He points out that time allows deep regoliths to develop; the reduction of resistant masses; the development of duricrusts; organic evolution, the impact of earlier events on subsequent change and the development of anomalous river patterns. In doing so he supports the propositions arising from Bremer.

His well-reasoned and illustrated argument also fulfils one of the aims of the original paper (p. 3, Ten C's) which asked, Have we retained or confirmed any of our earlier statements? It seems obvious to me having heard Twidale's debate that we still have much to learn from the Davisian trinity (*pace* Yatsu and Tricart) perhaps in the following form:

Proposition D: All landforms are a function of geomorphological processes acting on geological structures and materials over the time available.

This is at once simple, straightforward and pertinent. That it was enunciated almost a century ago does not render it invalid. Nor does its authorship put it out of court. I have been accused in the debate (see Yatsu and Tricart) of being neo-Davisian both because I have proposed a new structure for the discipline and because I have accepted some of Davis' long-held propositions. The first is silly. You might as well label me a neo-Gilbertian, a Schummian a Scheideggerian or even a Bakerian because they too have stated propositions and I would accept and use much of their work until something better comes along. The second is a delight because I am constantly finding some Davisian ideas of assistance in my attempts to understand.

Twidale's last point is to review, in a refreshing manner, the role of water. I have no points of rebuttal and accept his proposals. In fact I would make it a "cardinal" point because it is such a general and long-held belief in the subject (e.g. the normal cycle of erosion!). For example, it is an early proposition of my own that "Water like Heineken reaches parts that other agents cannot reach" (Brunsden 1972).

I am therefore happy to accept Twidale's proposition:

Proposition E: Water is the critical factor in landform evolution on earth (Twidale 1996, this volume).

May I thank you for such a thoughtful debate and for the great compliment you pay me in the opening words of your summary. In return I can say that I have long been a Twidalian.

Eiju Yatsu

Professor Yatsu, who attended the conference in order to take part in the debate, even though he had recently been seriously ill, made several important comments. I have already dealt with the use of the word "commandment" and with the objectives of the paper. More substantively, he is concerned with the use of the word "catastrophe". Tricart and Baker also do not like this. I agree with them that it is a serious issue. Yatsu, unlike Baker, accepts my definitions of the word but not the word itself because it has historical meanings which are unacceptable (to any of us?) and because it might lend support to the creationist movement. I support his viewpoint but I am not sure what to do about it.

There is a point in every system when an event can occur which overtops the frequency–magnitude scale of the processes **which create and maintain the system**, and which "thoroughly overturns it" so that "a new system is created". This I have called a geocatastrophe. I am happy to accept another word; perhaps geocataclysm would do?

To understand this type of event is not a problem if the event is a new process caused by a change in the control specifications of the system (e.g. if a lava flow issues for the first time from a vent and flows down the nearest valley thus forcing the river to flow somewhere else and to create a new river system.)

The problem arises when the controls do not alter but a system-altering event still occurs. The point is that neither the controls nor the type of process changes but the system behaviour changes, it self-destructs or — as I have described it — is wiped out. I need to think about it a lot more. I therefore leave the subject open to debate while the thought processes develop and thank Yatsu for the challenge.

Jean F. Tricart

What a pity Jean Tricart was not in either Frankfurt to hear the paper or in Hamilton to take part in the debate. I am sure that most of the points he raises are based on a misunderstanding of what I was trying to say? Since others have complained that I have used abstruse language I have to accept that this is my fault. Again I apologize. What I will not accept is the attack on my so-called Davisian stance. It is true that I was taught in the Davisian tradition and I am intensely proud of having studied with S.W. Wooldridge but I have also learnt from Tricart, Budel, Schumm, Leopold, Kirkby, Thornes and all that I have read in the last 40 years. The paper given at Frankfurt is the synthesis. As I have said, I see nothing wrong with some of the statements of Davis and see no reason to reject them just because he said them. If they smell of Davis then they are the sweeter for it.

By and large, however, I agree with Jean Tricart's approach and with his debate. The debate and his life's work are fine contributions spoilt only by his obsessions. Actually there are few points of disagreement between us. A careful reading of what I wrote alongside the critique will show that in most cases I have been misunderstood and also that I have been much influenced by Tricart himself.

As examples:

1. Tricart complains that I give lists of incompatible examples. They were meant to be different examples. If the text is read carefully there would be no confusion as to my intention.

2. Tricart comments on the contradictions between Propositions 4 and 5 suggesting that constant processes are incompatible with perturbations. Not so. The idea expressed is that a system which is operating with a set of steady conditions will settle down to a steady behaviour and will probably produce a regular suite of landforms. This is the basis of steady-state or dynamic equilibrium concepts in modern geomorphology. All I do in Proposition 5 is to recognize that such steady conditions are short-lived and that the system will sometimes be disturbed. It will react to this, respond (relax) and settle down again to the new regime. This is perfectly compatible with nature. For example Jean Tricart lives a fairly sober life with a constant behaviour pattern and a reasonable geometry. This does not mean that he has had no perturbations to his system nor that he has never had to get in shape again.
3. He dislikes my use of very general models and denudation rates (fig. 3 in Brunsden 1990) and the idea of sea level as a control. I agree with the former, although the figure was trying to make a point which was not dependent on precise data. His rejection of sea level because "it has the nasty smell of the Davisian cycle of erosion" is ignored. I suggest we all take a look at those sea-level curves again and ask if we can reject them as a fundamental fact?
4. Tricart says he does not understand my use of catastrophes. I am not sure it is right either and I share with him the problem that I am at the beginning of my thought process on the subject. But he has done us a service by pointing out that the caption for my fig. 8 is incorrect and makes the diagram absurd. Figure 8 from Brunsden (1990) is hereby reproduced as Figure 2.3 and a new corrected caption inserted.

   My point is that we tend to adopt the idea that a catastrophe only occurs on the human time scale and thus has a very short duration. The fact is that if we consider the time scale of earth history of 4000 million years (rather than a human time scale of 100 years) then orogenies or transgressions assume something of the character of catastrophe and, in those time terms are also of short duration.
5. Tricart returns to his Davisian obsession in his consideration of Propositions 4 and 5. The suggestion that a rapid response of a system (i.e. a steep curve on my fig. 2 in Brunsden 1990) is Davisian is absurd. Nevertheless, even if it was something Davis has said, it would only show that Davis was correct in some things. Tricart seems to have missed the point that the Taiwan curve was based on real data!

   He complains that relaxation curves are Davisian. Has he never heard of an exponential curve? Relaxation can be positive as well as negative. He should also note that the time scale is not a Davisian 100 million years but any time scale and all events. Actually my fig. 4 is the key as we attempt to use our short-term process data to understand long-term landscape evolution. I will not allow him to reject it just because the curve has a pseudo-Davisian shape.
6. His last point has the same flavour or at least "intense smell of Davisian peneplains". At no time have I used the word and I do not imply that at the end a plain will be produced. The argument is that there are many mechanisms which prevent change. Only rock resistance is Davisian. To help Tricart I have elaborated on this in a paper in honour of Hanna Bremer (Brunsden 1993). Twidale and Bremer have discussed the point in this volume and I have agreed with them. I stand by Proposition 10, it may even be the most important.

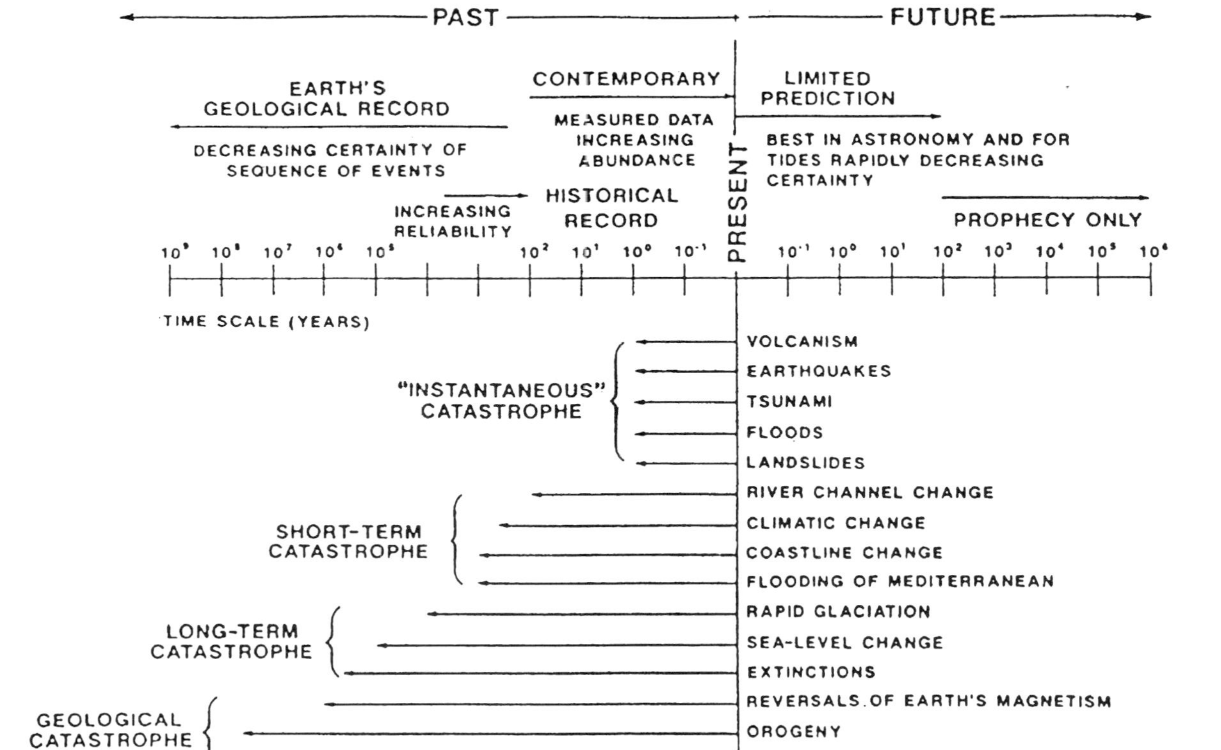

FIGURE 2.3 A classification of geocatastrophes (from Brunsden 1990, fig. 8). The lengths of the arrow represent the approximate time scale within which different types of "catastrophè" can be thought of as happening very quickly. Thus, for example, an orogeny taking 10 million years will be very sudden and "catastrophic" in character if viewed over a 10^9–10^{10} year period. The term ice age in the diagram is meant to represent the four to five times in earth history when "glacia" conditions have been recorded in the full earth time available, not one glacial lasting only *c.* 30 000 years. Jean Tricart is thanked for drawing our attention to the incorrect caption in fig. 8 (Brunsden 1990) which makes the diagram absurd

REFERENCES

Baker, V. and Twidale, C.R. 1991. The reenchantment of geomorphology. *Geomorphology*, **4**, 73–100.

Bremer, H. 1989. *Allgemeine Geomorphologie*. Gebruder Borntraeger, Stuttgart.

Brunsden, D. 1972. Weathering. In Embleton, C. and Thornes, J.B. (eds), *Processes in Geomorphology*. Methuen, London, pp 73–129.

Brunsden, D. 1993. The persistence of landforms. *Z. Geomorph. N.F. Suppl.-Bd.*, **93**, 13–28.

Brunsden, D. and Thornes, J.B. 1979. Landscape sensitivity and change. *Trans. Inst. Brit. Geogr.*, **4**, 463–484.

Budel, J. 1982. *Climatic Geomorphology* (translated by L. Fischer and D. Busche). Princeton University Press.

Chorley, R.J. and Haggett, P. 1965. *Frontiers in Geographical Teaching*. Methuen, London.

Davis, W.M. 1922. Biographical memoir of Grove Karl Gilbert, 1843–1918. *National Academy of Sciences, Biographical Memoirs*, **21**(5), Washington.

Griffiths, G.A. 1993. Estimation of landform life expectancy. *Geology*, **21**, 403–406.

Haines-Young, R. and Petch, J. 1986. *Physical Geography: Its Nature and Methods*. Harper and Row, London.

McSaveney, M. and Griffiths, G.A. 1988. A general theory for frequency distribution of age and lifetime of steepland elements formed by physical weathering. Unpublished personal communications available from the authors. NZ Geological Survey, Christchurch, NZ.

Popper, K. 1972. *The Logic of Scientific Discovery*, 6th revised impression. Hutchinson, London.

Yatsu, E. 1992. To make geomorphology more scientific. *Trans. Japanese Geom. Union*, **13**(2), 87–124.

3

The Point of Modelling Geomorphological Systems

FRANK AHNERT

Geographisches Institut der RWTH, Aachen, Germany

INTRODUCTION

In recent years there has been a considerable growth of interest in the role of modelling in geomorphology. Several books and numerous journal articles have appeared on the subject and entire symposia have been devoted to it. The term "model" is now widely used and to some extent has come to mean different things to different people. Yet terms are words that stand for concepts, and agreement about their purpose and meaning is essential. This paper attempts to identify the concept behind the term and to discuss its purpose by means of a few examples. Based mainly on three decades of personal experience with quantitative modelling, the paper expresses essentially a personal view, but perhaps it can nevertheless provide a few useful thoughts for reflection.

At first sight the new emphasis on models may seem like just another research fashion. It actually is more, namely a growing recognition that models are necessary as the formalized expression of geomorphological research methodology and of geomorphological thinking in general.

While the present emphasis on modelling may be new, the use of models is not. In fact, it is as old as geomorphology itself. In the past, Gilbert (1877) and Davis (1899), and also W. Penck (1924), Baulig (1940), L. King (1957), Büdel (1963) and others have interpreted landforms and landform development in terms of model concepts, even though they did not usually call their interpretations models. Their work has created the basis for the quantitative modelling that followed (after early attempts by Lehmann (1933) and by Bakker and Le Heux (1947)), from 1960 onward (cf. Scheidegger 1961; Young 1963; Ahnert 1964, 1966; Gossmann 1970; Kirkby 1971). The significance of their impact has been assessed by Young (1972) and more recently by Thorn (1988), while Anderson and Sambles (1988) have presented a comparative analysis of the quantitative modelling approaches of the 1960s, 1970s and 1980s.

The need for models is not confined to geomorphology. Rather, it has its roots in the basic structure of scientific thinking.

Geomorphology Sans Frontières. Edited by S. Brian McCann and Derek C. Ford.

All science begins with *perception*. Systematically organized, the perception becomes *observation*. Comparison of observed traits leads to *abstraction*, i.e. to the selection of those attributes that are considered essential and the deliberate omission of others that are not.

The attributes that are retained become criteria for the *classification* of the observed phenomena. The result of the classification is the identification of *types*. These types are no longer the reality that was originally perceived. Instead, they are man-made abstracted representations of that reality. In other words, they are *models of reality*.

KINDS OF MODELS

All general scientific statements or laws concern groups or types of phenomena, their properties and their interrelationships. Thus, they are statements about models. In keeping with the terminology of Windelband (1894) and Rickert (1902), two philosophers who taught in Heidelberg at the beginning of this century, such general models are called *nomothetic models* because they are part of the procedure to establish general rules or laws. The counterpart of nomothetic models are *idiographic models*, i.e. models which represent a particular individual case. A nomothetic model can be converted into an idiographic one by adding local specifying attributes to the more abstract general type.

Models derived inductively from field data are *empirical models*. This paper is devoted primarily to *theoretical models*. They are designed deductively from general principles, but these also are based to a large extent on the results of prior empirical research.

All geomorphological models are representations of systems. Their components are linked by functional relationships. All these models are therefore *functional models*. The systems terminology introduced into geomorphology by Chorley and Kennedy (1971) provides useful distinction of different kinds of such models (Figure 3.1).

Static models have components relating to form and/or material, but not to processes. They therefore do not describe change and do not contain time. Chorley and Kennedy (1971) have called systems of this kind "morphological systems", but in the context of geomorphology the term "static" instead of "morphological" helps to avoid possible confusion. An example of an empirical static model is the functional relationship between channel gradient and bedload grain size of a stream.

In *process models* time is an essential ingredient, especially as a means to express process rates, intensities and budgets. The process models represent what Chorley and Kennedy have called "cascading systems". The reference to processes seems better because "cascades" flow only in one direction while process systems frequently contain feedbacks. Sediment rating curves are examples of empirical process models.

Process response models, finally, combine the properties of static models and process models and therefore allow a comprehensive representation of form, material and process, of their functional interactions and of their change over time.

Among the nomothetic purposes of process response models are

- the analysis of functional relationships between particular components of the overall system, and
- the testing of hypotheses concerning the development of certain landform types.

Among the idiographic purposes are

Static models

Components: landforms and / or landform properties
materials and / or material properties

Functional relationships:
landforms = f (landforms)
materials = f (materials)
landforms = f (materials) or vice versa

Process models

Components: processes, process rates or
other process properties

Functional relationships:
process A = f (process B)
(with possible feedbacks)

Process response models

Components: landforms and /or landform properties
materials and /or material properties
processes and /or process properties

Functional relationships:

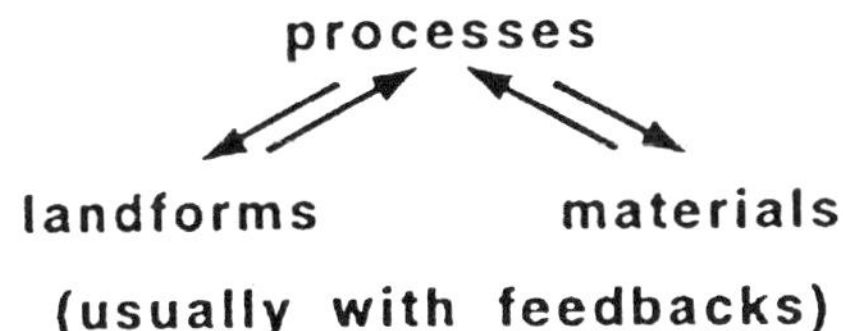

(usually with feedbacks)

FIGURE 3.1 Functional geomorphological model types

— the fitting of the general nomothetic model to the investigation of a local problem by including specific local conditions and factors, and
— the simulation of the evolution of particular individual landforms, landform assemblages, or landform sequences.

Idiographic models may also become the basis for predictions. At present, the capability to predict is limited to few subject matters and to very short time spans, and is probabilistic rather than deterministic. *Model experiments* have a similar purpose to physical laboratory experiments, namely, to determine the effects of a change of individual model components upon other components and upon the system as a whole.

Because of their particular advantages and disadvantages, models, laboratory experiments and field research complement each other and are ultimately interdependent. As Brooks *et al.* (1993) have pointed out:

> Field measurements must invoke the ergodic principle, laboratory experiments introduce problems of scale representation, and mathematical models require careful construction and calibration ... Frequently a combination of approaches represents the most successful means of elucidating both the nature and rate of long-term ... development.

A similar complementarity exists between empirical and theoretical geomorphology in a more general sense. Neither can advance its research effectively without the other. Empirical geomorphology provides the observational evidence from which theoretical considerations may be developed; in addition, it serves to confirm or reject the validity of theoretical concepts. Theoretical geomorphology, on the other hand, combines the available empirical knowledge with the application of general physical principles in order to arrive at the formulation of theories which in turn can suggest goals for further empirical research.

Figure 3.2 shows a generalized diagram of the relationships and interactions between the major components that participate in the development of fluvial landforms. Note the numerous feedbacks.

Even within the realm of this rather limited system, very many different models have been developed (see, for example, Ahnert 1976a, 1987a; Woldenberg 1985; Kirkby 1987; Anderson 1988; Phillips and Renwick 1992).

Apart from the eksystemic energy inputs, the system components are forms or form parameters (such as slopes and channel beds), materials or material parameters (such as rocks and structures, regolith, or stream load), and processes which transfer material: weathering transfers bedrock material to the regolith, denudation transfers the regolith downslope, and the stream flow transports it away, while crustal movements "transfer" parts of the crust upwards or downwards and thus produce the elevation differences that provide the necessary potential energy.

Each transfer can be expressed by an equation which is as far as possible derived from empirical experience and which must be in agreement with physical laws. Indeed, much of geomorphology is applied physics. All of geomorphology is, or should be, also applied common sense. It dictates which aspects of physics are to be included and which are to be left out. There is not much of a point, for example, in analysing the total energy budget of the system, because only a small part of the energy is actually converted into morphological work. Most of the total energy input passes through the system without causing any change of forms or materials. Landform development is first and foremost the result of transfers of materials from one location to another. Therefore the mass budget is more immediately important in geomorphology than the energy budget.

Within the larger scheme, the functional interactions between forms, materials and processes at a surface point take place as shown in Figure 3.3. If the land surface is represented as a field of grid points such as this one, and if each grid point is linked to its neighbors by these functional relationships, then the whole becomes a process-response model of the land surface.

This is the idea behind the design of SLOP3D. It is a rather unsophisticated but also fairly effective model of landform development which serves here as an example. The

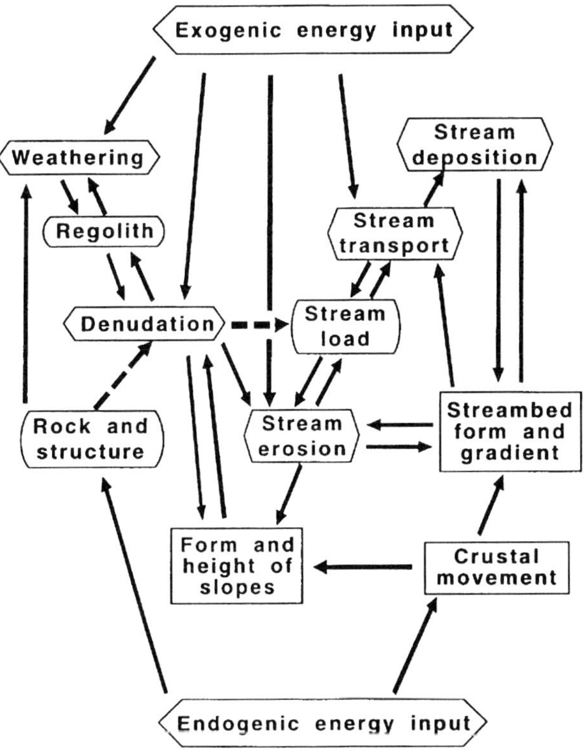

FIGURE 3.2 Components and interactions of the fluvial landform system (schematic)

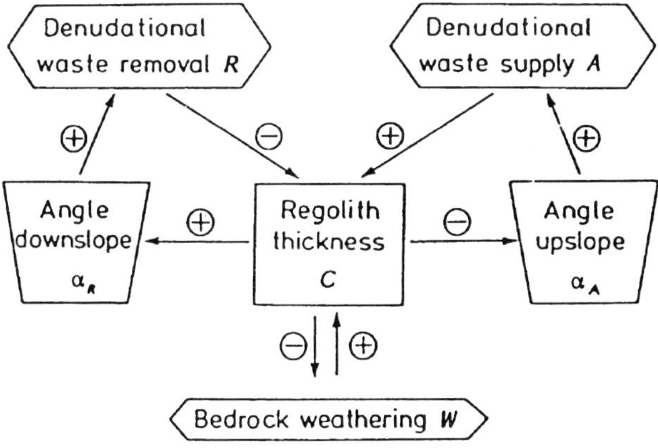

FIGURE 3.3 Functional interaction of system components at a surface point (after Ahnert 1987b)

basic structure of this model is already several decades old and has been explained in various publications (e.g. Ahnert 1971, 1973, 1976b, 1987b, 1988). Figure 3.4 shows the major components of the model program in a condensed manner.

EXAMPLES OF MODEL APPLICATION

Over the years, SLOP3D has been used in many ways for both nomothetic and idiographic models, with special emphasis upon theoretical model experiments. Many of these have been published (cf. references to Ahnert above) and therefore need not be repeated here.

Instead, two new model series may illustrate some uses of SLOP3D and demonstrate the point of modelling as well. Each series leads from the consideration of a basic concept over a test by means of profile versions of the model to the simulation of landform development in three dimensions. The first series concerns the development of cuestas and structural benches, the second concerns the development of inselbergs on lithologically and structurally homogeneous rocks.

According to conventional wisdom cuestas develop on slightly dipping strata, with a resistant bed overlying a weak bed and the contact between the two exposed laterally.

If this concept is translated into a series of profile models, there develops indeed a sort of scarp. It looks, however, quite different from most real cuesta scarps: from the dip slope towards the scarp crest, the caprock thins out into a knife-like edge and gets worn thin generally (Figure 3.5, no. 1–3).

The mere identification of a greater overall resistance of the caprock obviously does not suffice to explain the cuesta development. How then must the notion of caprock resistance be modified so that a reasonable semblance of cuesta development can be obtained?

Field observation shows that the caprock is not uniformly resistant in all exposure directions. Rather, it is most resistant on the exposed bedding plane, where much of the rainwater percolates into the network of joints and therefore is not available for surface denudation. On the scarp slope, however, the joints are exposed in a steeply sloping surface. Lateral pressure release tends to open them and, together with weathering, to encourage joint block separation and, eventually, rockfalls and landslides. The weakest zone of the "resistant" caprock, finally, lies at the contact with the underlying "weak" rock, where sapping by springs and seepage undercuts the rock and thereby steepens the slope above.

Thus it is necessary to distinguish at least three different resistance zones on the caprock. This interpretation is now applied, as the only change, to the profiles no. 4–7 in Figure 3.5. The results correspond much better to the appearance of cuestas in the field.

Of course such an interpretation seems very obvious as soon as one thinks of these points. But without the simple model experiment, this obvious thought might not have arisen. It can now be applied to the three-dimensional version of SLOP3D.

In the first example (Figure 3.6), the initial surface is horizontal with some minor random variations that cause local convergences and divergences of overland flow. A resistant bed dips from front to rear. Stream incision has been programmed to take place continuously at both ends of the model block. Weathering converts the bedrock into regolith, and denudation is by surface wash which in this model can only remove regolith material.

After 10 time units of the model run, the front slope extends downward across the contact between resistant and weak rock, and is already noticeably steeper that the slope at the rear of the model. At time T=40 (arbitrary computer time units the stream incision

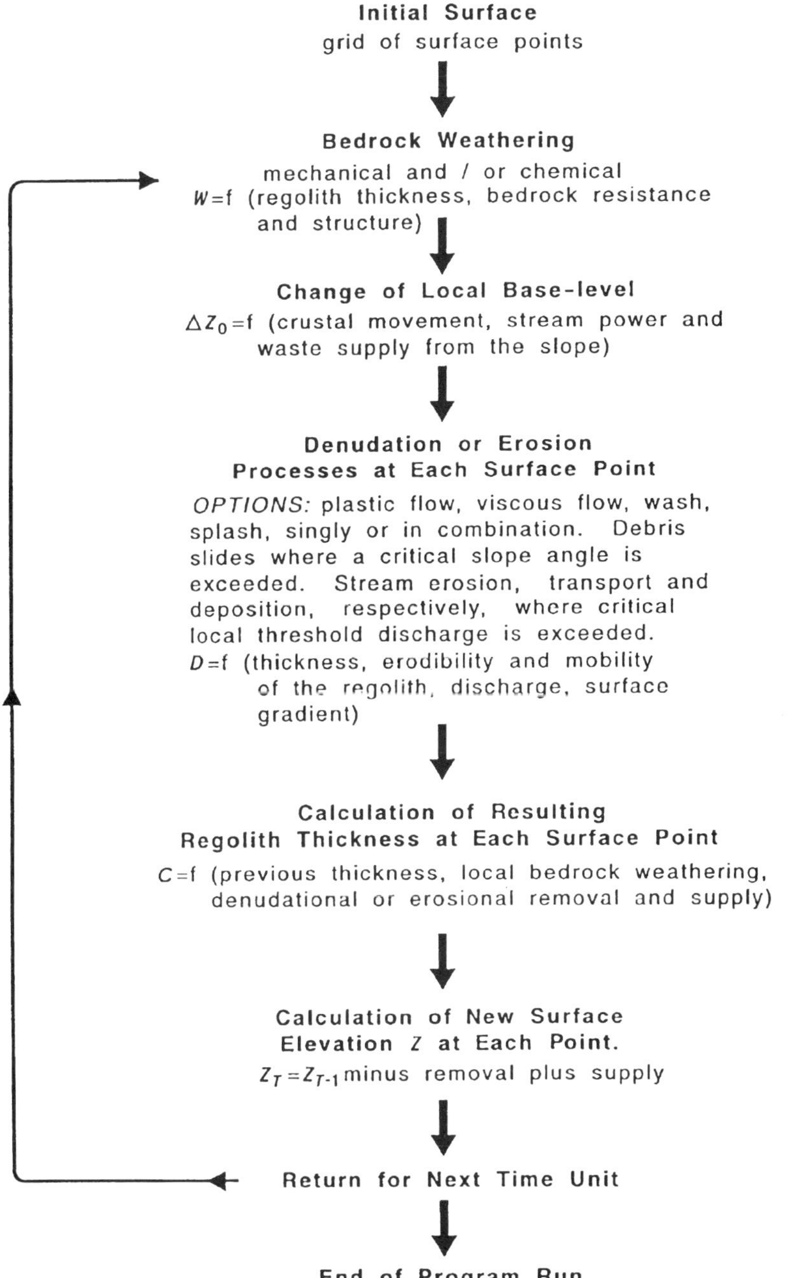

FIGURE 3.4 Simplified design structure of SLOP3D-type landform development models

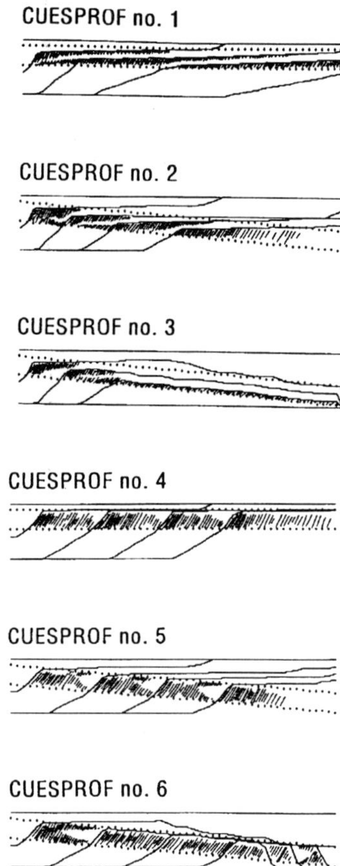

FIGURE 3.5 Profile models of cuesta development. Dotted lines mark the stratigraphic boundary of the resistant bed. In profiles no. 1–3 the resistance of this bed is uniform, in profiles no. 4–6 it is directionally differentiated

at the rear has also cut through the resistant bed, so that a rear scarp (an "Achterstufe" in the terminology of Mortensen (1953)) is developed. The resistant bed is progressively consumed by the retreat of both the front and the back slope.

In the model shown in Figure 3.7 the surface runoff can also cut into bedrock if the discharge exceeds a predetermined threshold value. This leads to the incision of channels and, consequently, to valley formation.

This model already comes quite close to the appearance of natural cuesta landforms. Of special interest is the contrast between the relatively straight front scarp and the deeply indented rear scarp. The vertical scale of the model shows that the flat summit level from time *T*=70 onward is not a remnant of the initial surface but the undissected part of the dip slope.

So far, the models have been of the nomothetic kind. By adding information about rocks and bedding structure of a particular locality, the nomothetic model is converted

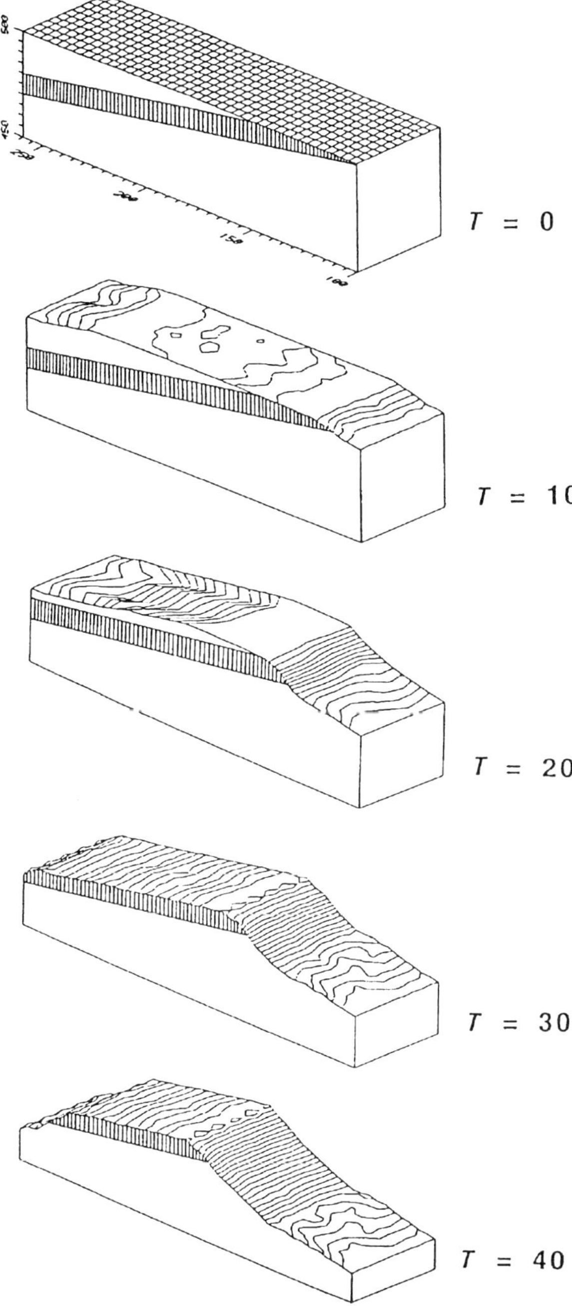

Figure 3.6 A simple three-dimensional model (HAMCUES No. 3) of slope development. The initial surface contains a very slight, randomly varied relief which causes local divergence or convergence, respectively, of overland flow and thus of denudation. The resistant layer is indicated by hachuring

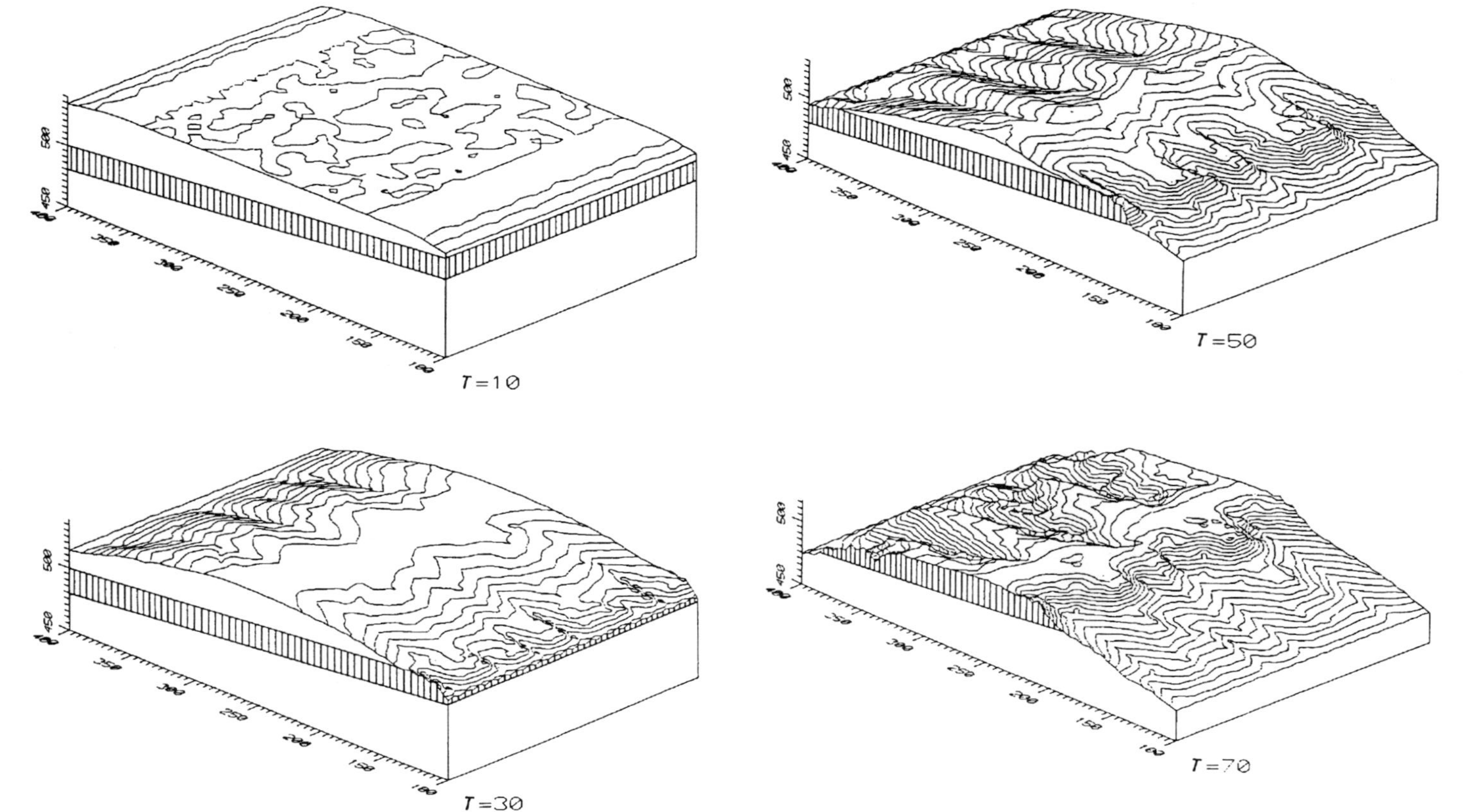

FIGURE 3.7 *For caption see opposite*

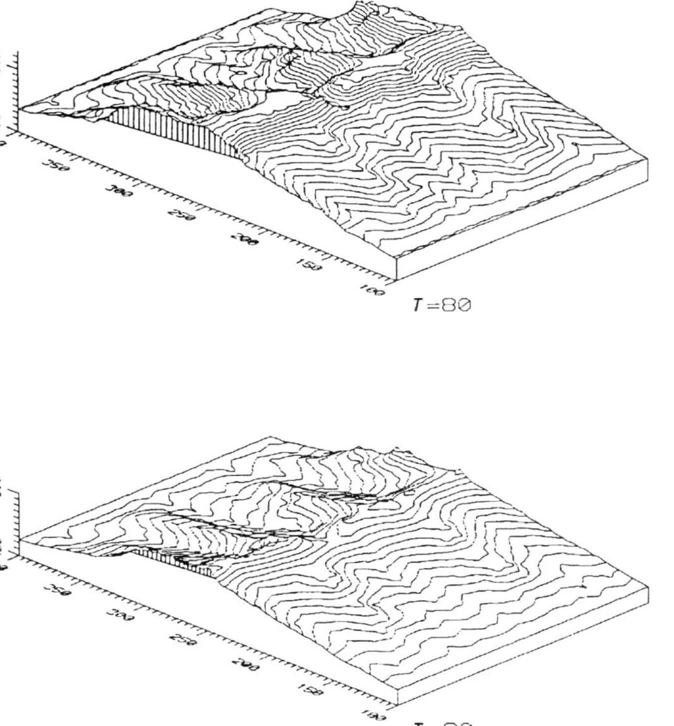

FIGURE 3.7 (*and opposite*) Three-dimensional model (HAMCUES No. 5) of cuesta development. The processes are the same as in Figure 6, but with the additional erosion of bedrock where the runoff discharge exceeds a pre-set threshold

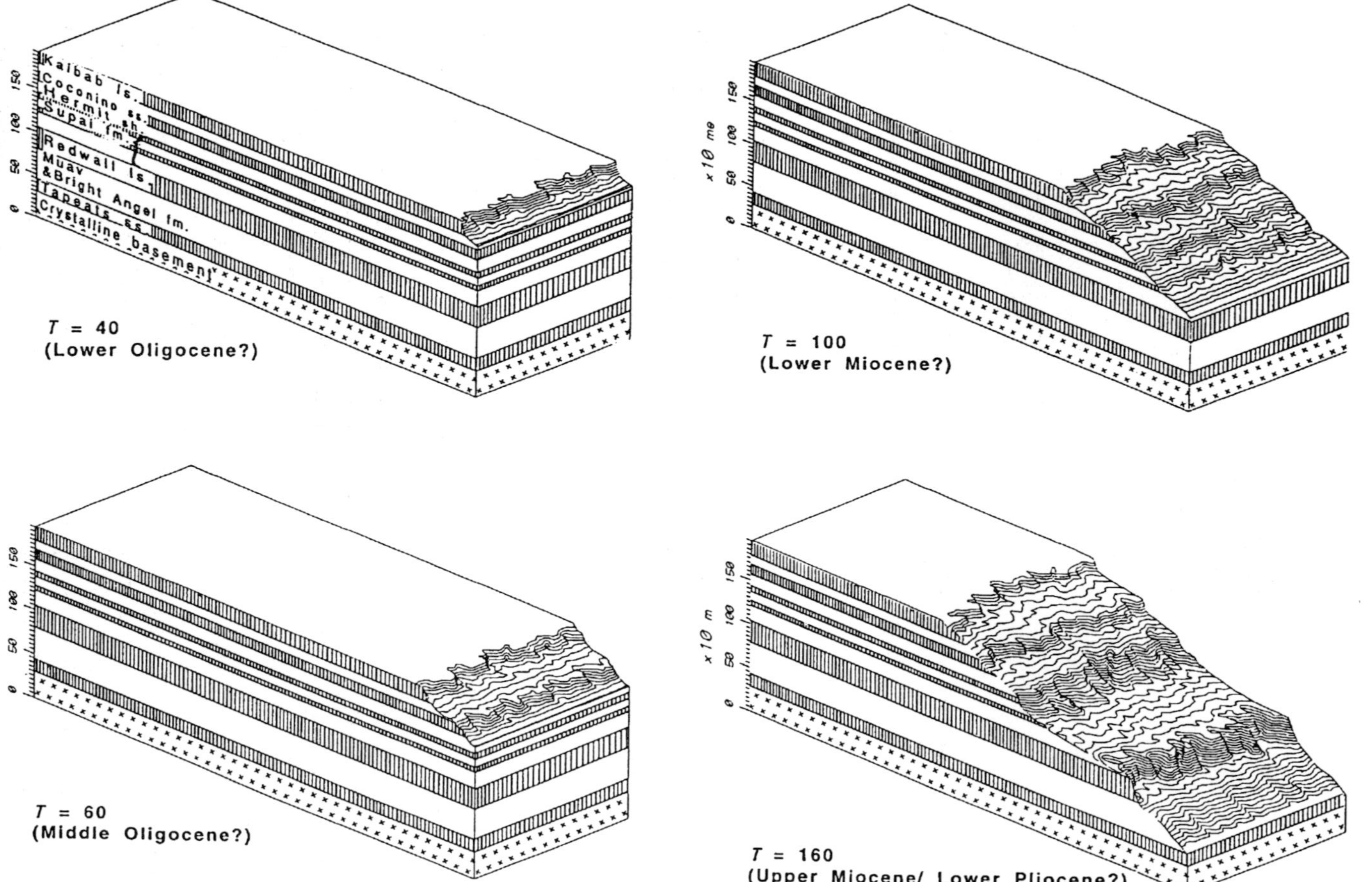

FIGURE 3.8 *For caption see opposite*

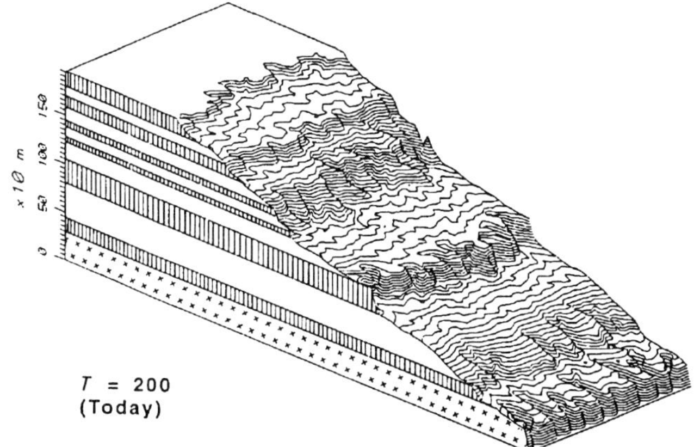

FIGURE 3.8 *(and opposite)* Three-dimensional idiographic model of the development of the Grand Canyon of the Colorado River, Arizona, USA, since the end of the Eocene (Grand Canyon Model No. 16). Only one valleyside is shown. The vertical erosion of the Colorado River is assumed to have taken place continuously and at a constant rate. Estimates of the geological ages of the different phases of development are therefore only crude approximations

into an idiographic one. The development of the Grand Canyon with its structural benches may serve as an example of such a conversion (Figure 3.8).

The only significant change from Figure 3.7 is the incorporation of the approximately horizontal stratigraphic sequence of the major resistant and weak rock formations that occur in the Grand Canyon region. The strata differ in terms of their weathering rate; for the scarp-forming beds, the weathering rate is differentiated directionally as explained above.

For lack of better information, the vertical erosion by the Colorado River is assumed to have taken place continuously and at a constant rate since the end of the Eocene. Each time unit of the model would therefore represent approximately 200 000 years of real time. Accordingly, the Kaibab limestone scarp would have begun its retreat during the lower Oligocene, the Coconino sandstone scarp in the middle Oligocene. The series of resistant and weak beds that make up the Supai formation were cut through by the river presumably in the upper Oligocene and lower Miocene, and the Redwall limestone during the middle and early upper Miocene. From the middle Pliocene to the present, the scarp of the Tapeats sandstone with the Tonto platform developed and the "Granite Gorge" was incised.

The mean slope from the canyon rim to the river is about 26%, which is very similar to the mean slope in many places of the South Rim of the actual Grand Canyon. The position of the individual structural benches in the overall slope at time T=200 also represents the present position of these benches in the Grand Canyon reasonably well. If the times and degree of variations in the rate of downcutting by the Colorado River were known and could be entered in the model, the fit might become even better than with the present assumption of uniform downcutting. This is not the first evolutionary model of the Grand Canyon (cf. Pollack 1969; Cunningham and Griba 1973; Aronsson and Linde 1982), but it bears a greater resemblance to its real landforms.

The second example of model applications concerns the development of inselbergs. These landforms usually have been described as a result of differences in resistance, due either to differences in the mineralogical composition of the rocks or to structural differences, especially to variations in the spacing of joints, much in the way Linton (1955) has explained the origin of tors.

Such explanations are certainly correct in many cases, but not always. In the Highland of Machakos, Kenya, for example, the large inselberg Kilungu consists of muscovite gneiss and parts of the pediment surrounding it of granite gneiss, while about 10 km north of it at the Inselberg Mbooni the opposite is true (cf. Ahnert 1982). In petrographically homogeneous areas, variations of joint spacing are difficult to verify if, as is most often the case, the pediment that surrounds the inselberg is regolith-covered. Already Jessen (1936) has postulated that inselberg formation should be possible in the tropics without lithological or structural influences if the bedrock gets exposed and therefore is less subjected to chemical weathering than the rock at the base of a moisture-storing regolith. This concept of "divergent weathering" (Bremer 1971, 1973) is related to the hypothesis concerning the development of "etchplains" in the tropics (cf. Thomas 1974, pp. 240–244). In a later paper Thomas (1978) has presented a comprehensive review of the various types of inselbergs and their possible modes of origin.

The model SLOP3D can be used to test the feasibility of Jessen's interpretation. The chemical weathering function of the model program has been designed in such a way that the maximum rate of bedrock weathering is reached under a regolith cover of predetermined thickness C_c which is, for the local climatic conditions, great enough to ensure a continuous presence of moisture at the bedrock/regolith interface. At lesser thicknesses the regolith dries out at times, so that bedrock weathering is interrupted. At greater thicknesses, CO_2 and other weathering agents are increasingly consumed by further weathering within the regolith, and therefore are not available for the weathering of the bedrock.

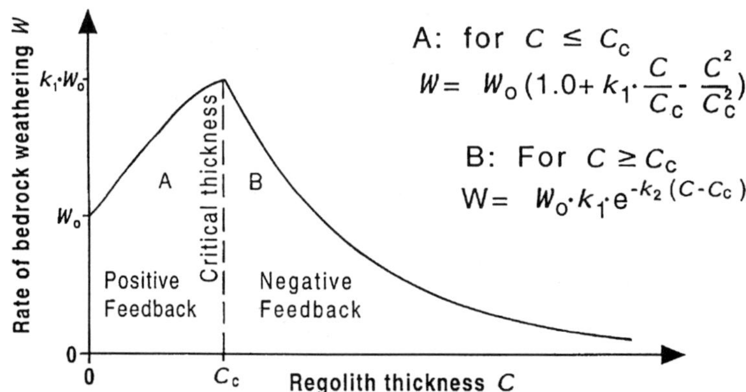

FIGURE 3.9 The chemical weathering function of SLOP3D (after Ahnert 1977)

Along the two limbs of the weathering curve on either side of the maximum, the weathering rate reacts differently to changes of the regolith thickness by denudation. On the descending limb (zone B in Figure 3.9), a thinning of the regolith cover causes an increase in the rate of weathering, a thickening causes weathering to decrease. This is a negative feedback which eventually leads, if the denudation rate remains constant, to a dynamic equilibrium between denudation as the forcing variable and weathering as the responding variable.

On the rising limb of the weathering curve, however, there is a positive feedback. Thickening of the regolith causes weathering to increase, which makes the regolith even thicker, and thinning causes weathering to decrease, which makes the regolith even thinner. The realm of positive feedback is limited at both ends by boundary conditions: If the regolith becomes progressively thicker, it eventually reaches the thickness C_c at which weathering is most intensive and at which the negative feedback sets in. On the other hand, progressive thinning of the regolith eventually leads to the baring of the bedrock. Zone A of the chemical weathering function therefore contains these two opposing or, in Bremer's terms, divergent tendencies, namely, that an increase of regolith

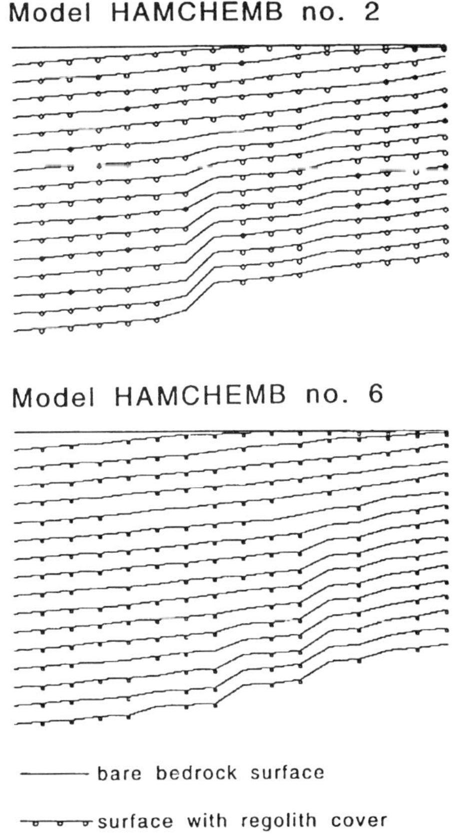

FIGURE 3.10 Profile models with divergent weathering

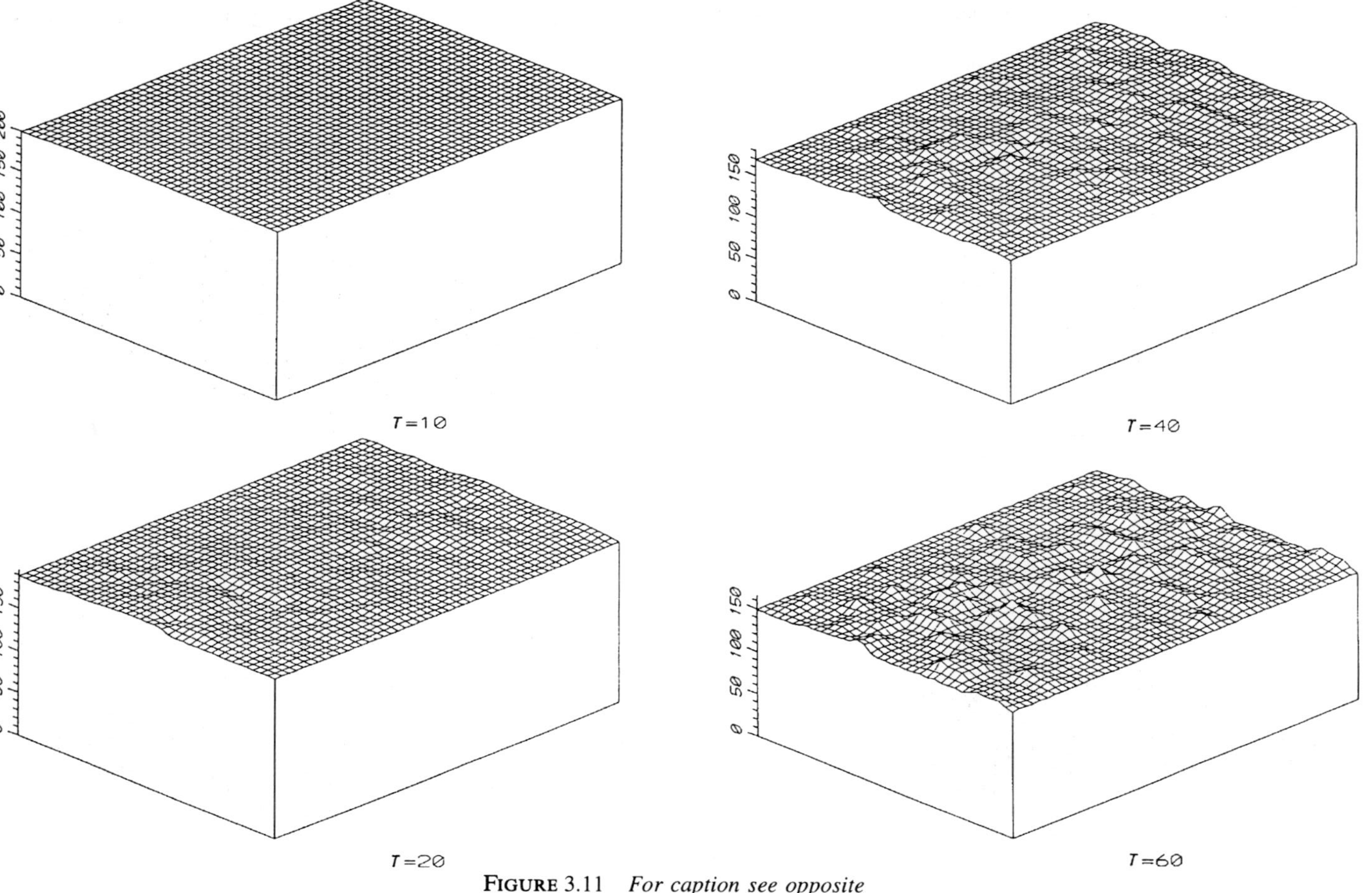

FIGURE 3.11 *For caption see opposite*

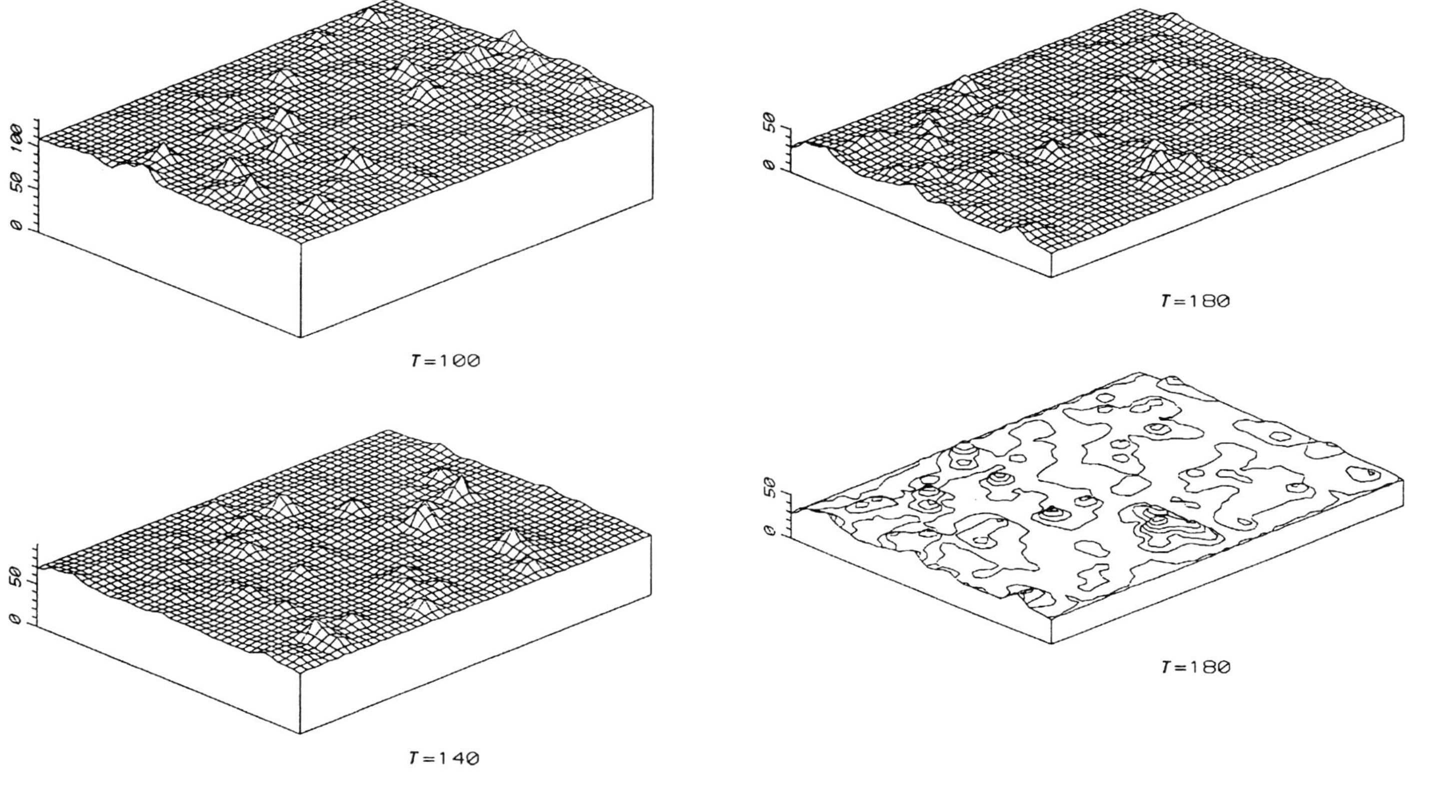

FIGURE 3.11 (*and opposite*) Three-dimensional model (HAMINS No. 11) of inselberg development on petrographically and structurally homogeneous rock

thickness causes further thickening, while a decrease causes further thinning until the bedrock is bare. It stands to reason that in regions with dry seasons C_c must be greater than in permanently humid regions. The greater C_c, the greater is the probability that the conditions of regolith thickness, weathering rate and denundation rate lie in zone A of Figure 3.9, and the greater therefore is the probability that divergent weathering does indeed occur. This may provide the answer to the question why inselbergs are more frequent in the subhumid and semi-arid tropics than elsewhere. Direct empirical evidence of the two divergent tendencies of chemical weathering in granitic semi-arid to subhumid environments has been supplied recently by Twidale (1993, p. 196) with his observations of rock levees in Zimbabwe and of "rock doughnuts" in central Texas.

As a first step of modelling, these functional relationships are applied to profile models under conditions of continuous downcutting of the local base-level (Figure 3.10). The rate of denudation, which is normally programmed in SLOP3D as a function of local discharge and gradient (cf. Ahnert 1977), is varied randomly within narrow limits. Successive high-intensity denudation events can shift the regolith thickness into the zone of positive feedback where it will either recover or become zero. In this way, greatly differing rates of regolith production by bedrock weathering are possible on an entirely homogeneous bedrock.

The presence of regolith is indicated in Figure 3.10 by the small circles immediately below the profile lines. Repeatedly there occur bare spots at which, however, the regolith recovers. At other places, the differential weathering causes the bare rock areas to become less lowered by denudation than the regolith-covered points downslope from them, with the result that a knickpoint develops. The steeper gradient at these points increases the denudation rate. This in turn helps to keep the point bare and thereby perpetuates the process of form differentiation.

A disadvantage of profile models is their linearity which allows transfers of material only along the profile, i.e. across the knickpoints on bare rocks rather than around them. Because of this artificial limitation the knickpoints automatically take on the function of local base-levels for the profile segments that lie above them. This difficulty is overcome in the three-dimensional version of the model (Figure 3.11).

The initial surface is flat with very minor random variations in height. Continuous fluvial incision takes place at the forward and rear edges of the model field. The bedrock is homogeneous throughout, and the denudation processes are the same as in the profile models of Figure 3.10. The vertical scale has been placed at the left rear corner of the model field in order to indicate the progressive lowering of the land surface as a whole.

After 40 time units of development, the local relief has increased in several places. Broad, gently sloping hills have formed, similar to shield inselbergs.

After 60 time units, these hills are more accentuated. They stay behind in the general lowering of the surface because their summit areas are either bare or only covered by a very thin regolith which keeps them in the positive feedback zone of the weathering function.

Eventually, the hills develop into azonal inselbergs. As the land surface is lowered further, some inselbergs disappear because the regolith in their summit area has become so thick that its further development is dominated by the negative feedback relationship which tends to diminish local differences in the weathering rate. On the other hand, new inselbergs begin to form elsewhere.

The flattening and disapearance of previous inselbergs due to negative feedback in some places, and the creation of new inselbergs due to positive feedbacks in other places combine to keep the overall relief and appearance of the model landscape in an approximate steady state, although there is no steady state locally at the sites of the individual inselbergs. The regional equilibrium between the rates of weathering, denudation and fluvial lowering is accompanied by varying local states of disequilibrium.

The last diagram of Figure 3.11 shows the model inselberg landscape of time T=180 by contour lines in order to make the relative heights of the various inselbergs better visible.

Seen from a lower angle of view, the model landscape perhaps also looks somewhat more realistic that from the higher angle (Figure 3.12). These diagrams are very reminiscent of the drawings which Jessen made of the inselberg landscapes in Angola about 60 years ago (Figure 3.13).

This model experiment shows indeed that inselberg development on homogeneous rock is possible. It implies also that even where they exist, petrographic or structural differences between an inselberg and the adjacent pediment do not automatically need to be accepted as the cause of inselberg formation. In other words, mere spatial association of a difference in form with a difference in lithology or structure is no sufficient proof for a causal relationship. The presence of such control can only be assumed safely where several inselbergs occur in a noticeable alignment with such inhomogeneities of rock type or structure.

CONCLUDING REMARKS

Three points might be made in conclusion. The first point concerns process geomorphology.

This paper has largely been confined to a discussion of process response models because they are most comprehensive and thus of the greatest general interest. On the

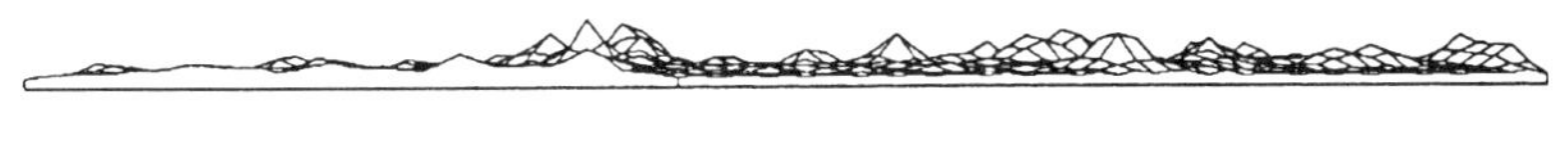

HAMINS no. 11, T=110

HAMINS no. 11, T=170

FIGURE 3.12 Low-angle views of the model inselberg landscape at times T=110 and T=170

Near Caibambo, Angola

Near Cubal, Angola

FIGURE 3.13 Drawings of inselberg landscapes in Angola (after Jessen 1936)

other hand, much modern research in geomorphology concentrates upon process mechanisms, process measurements and assessments of process magnitudes and frequencies. The preponderance of such themes is no doubt the result of a felt need to improve our knowledge in this field that has been much neglected in earlier decades of this century. It is also a consequence of the growing interest in applied geomorphology which is largely process-oriented.

These approaches are certainly useful and perhaps also necessary at present. In the long run, however, such a heavy emphasis on process geomorphology may cause a problem. Geomorphology is primarily the study of land*forms* and their development over time. Knowledge of processes is a requirement, but not the ultimate purpose. We must retain the ability to investigate long-term landform evolution as well as short-term landform changes in terms of process response systems, not just process systems. Otherwise geomorphology might eventually lose its identity as a recognizable discipline of the earth sciences. For example, civil engineers today measure and investigate earth surface processes also, sometimes very well and admirably. But we are not really civil engineers and they are not really geomorphologists.

The second point concerns the question of model design and model resolution. There are two fundamentally different ways to the design of a process response model.

One way is to start with the smallest identifiable system elements (stream pebbles, soil particles, raindrop sizes, individual plants, freeze/thaw cycles, rainfall and runoff events etc.), and to assemble these elements and their interrelationships much as one would assemble a mosaic out of many small stones.

Some might call this the white-box approach, because all components and their relationships are identified from the start. Actually, there is no such thing as a "white box", because every one of its elementary components is still a black box, however small it may be.

This approach has the disadvantage of going from the start into detail that may not really be required for investigating a particular problem. Also, since not all these many components are directly known from empirical measurements, many assumptions about their magnitudes and their interactions have to be included in the model. And the greater the number of assumptions, the less reliable is the model.

The other approach starts with a coarse representation of the system as a whole, in effect a big black box, and then divides it into still fairly coarse components which, with their interactions, can be quantitatively described. The subdivision of the larger system into smaller components continues as far as the problem to be modelled requires and the available knowledge allows.

Both approaches have been tried by various authors. In the design of SLOP3D the latter approach was chosen, for several reasons:

- from the beginning, the model is defined as a whole, with clear boundaries;
- eksystemic inputs into the model remain distinguished from ensystemic transfers and feedbacks;
- by dividing the model into components only as far as its purpose requires, the number of variables and of their functional interactions remains relatively small;
- the model can easily be modified for the investigation of particular questions or to simulate particular local conditions, by changing the parameters of some model components or by adding specifications to the overall model design.

This flexibility makes SLOP3D especially useful for carrying out theoretical model experiments.

The model can be fitted to any degree of spatial resolution by adjusting the distances between the surface grid points in conjunction with the spatial scale chosen. It can be fitted also to any degree of temporal resolution by modifying the magnitude of the time steps in relation to the chosen time scale, and it can be fitted to any required subject matter resolution by specifying such additional components and parameters as soil horizons, pore space, grain sizes, vegetation cover etc.

From this follows also the third point, which is a plea for simplicity in the face of seeming complexity. With apologies to Shakespeare, one can say "There is nothing either simple or complex in nature, but thinking makes it so."

This has been a problem throughout the history of the sciences, as shown, for example, by the model of the solar system of Ptolemy compared to that of Copernicus and Kepler. Their difference is not due to different empirical data, but to different thinking. A complex explanation is not automatically better than a simple one, even though it may be more impressive. Moreover, it can become an obstacle to further progress of understanding. As William of Occam wrote about seven centuries ago in a statement that sometimes is referred to as "Occam's razor": "Pluralitas non est ponenda sine necessitate".

Freely translated, this means: "Don't make matters unnecessarily complicated".

REFERENCES

Ahnert, F. 1964. Quantitative models of slope development as a function of waste cover thickness. *Abstracts of papers, 20th International Geographical Congress, London*, p. 118.

Ahnert, F. 1966. Zur Rolle der elektronischen Rechenmaschine und des mathematischen Modells in der Geomorphologie. *Geographische Zeitschrift*, **54**, 118–133.

Ahnert, F. 1971. A general and comprehensive theoretical model of slope profile development. *Univ. of Maryland Occasional Papers*, **1**, 95 pp.

Ahnert, F. 1973. COSLOP2 — a comprehensive model program for simulating slope profile development. *Geocom Programs*, **8**, 24 pp.

Ahnert, F. (ed.) 1976a. Quantitative slope models. *Zeitschrift für Geomorphologie, N.F., Suppl*, **25**, 168 pp.

Ahnert, F. 1976b. Brief description of a comprehensive three-dimensional model of landform development. *Zeitschrift für Geomorphologie, N.F., Suppl.*, **25**, 29–49.

Ahnert, F. 1977. Some comments on the quantitative formulation of geomorphological processes in a theoretical model. *Earth Surface Processes*, 2, 191–201.

Ahnert, F. 1982. Untersuchungen über das Morphoklima und die Morphologie des Inselberggebietes von Mahakos, Kenia. *Catena Supplement*, **2**, 1–72.

Ahnert, F. (ed.) 1987a. Geomorphological models — theoretical and empirical aspects. *Catena Supplement*, **10**, 210 pp.

Ahnert, F. 1987b. Process-response models of denudation at different spatial scales. *Catena Supplement*, **10**, 31–50.

Ahnert, F. 1987c. Approaches to dynamic equilibrium in theoretical simulations of slope development. *Earth Surface Processes and Landforms*, **12**, 3–15.

Ahnert, F. 1988. Modelling landform change. In Anderson, M.G. (ed.), *Modelling Geomorphological Systems*. Wiley, Chichester, pp. 375–400.

Anderson, M.G. (ed.) 1988. *Modelling Geomorphological Systems*. Wiley, Chichester, 458 pp.

Anderson, M.G. and Sambles, K.M. 1988. A review of the bases of geomorphological modelling. In Anderson, M.G. (ed.), *Modelling Geomorphological Systems*. Wiley, Chichester, pp. 1–32.

Aronsson, G. and Linde, K. 1982. Grand Canyon — a quantitative approach to the erosion and weathering of a stratified bedrock. *Earth Surface Processes and Landforms*, **7**, 589–599.

Bakker, P. and Le Heux, J.W.N 1947 and 1953. Theory of central rectilinear recession of slopes. *Koninklijk Nederlandse Akad. van Wetenschappen Proc*, **50**, (8 and 9), **54** (7 and 8).

Baulig, H. 1940. Le profil d'équilibre des versants. *Annales de Geographie*, **49**, 81–97.

Bremer, H. 1971. Flüsse, Flächen- und Stufenbildung in den feuchten Tropen. *Würzburger Geographische Arbeiten*, **35**, 194 pp.

Bremer, H. 1973. Grundsatzfragen der tropischen Morphologie, insbesondere der Flächenbildung. In *Geographische Zeitschrift, Supplement "Geographie heute — Einheit und Vielfalt" (Festschrift for Ernst Plewe)*. Franz Steiner Verlag, Wiesbaden, pp. 114–130.

Brooks, S.M., Richards, K.S. and Anderson, M.G. 1993. Approaches to the study of hillslope development due to mass movement. *Progress in Physical Geography*, **17**, 32–49.

Büdel, J. 1963. Klima-genetische Geomorphologie. *Geographische Rundschau*, **15**, 269–285.

Chorley, R.J. and Kennedy, B.A. 1971. *Physical Geography — A Systems Approach*. Prentice-Hall Int. Inc., London, 370 pp.

Cunningham, F.F. and Griba, W. 1973. A model of slope development and its application to the Grand Canyon, Arizona, U.S.A. *Zeitschrift für Geomorphologie N.F.*, **17**, 43–77.

Davis, W.M. 1899. The geographical cycle. *Geographical Journal*, **14**, 481–504.

Gilbert, G.K. 1877. *Report on the Geology of the Henry Mountains, Washington.* (2nd edition 1880.)

Gossmann, H. 1970. Theorien zur Hangentwicklung in verschiedenen Klimazonen. *Würzburger Geographische Arbeiten*, **31**, 146 pp.

Jessen, O. 1936. *Reisen und Forschungen in Angola*. Reimer, Berlin, 397 pp.

King, L.C. 1957. The uniformitarian nature of hillslopes. *Transactions of the Edinburgh Geological Society*, **17**, 81–102.

Kirkby, M.J. 1971. Hillslope process-response model based on the continuity equation. In

Brunsden, D.(ed.), *Slope Form and Process. IBG Special Publication 3*, 15–30.
Kirkby, M.J. 1987. General models of long-term slope evolution through mass movement. In Anderson, M.G. and Richards, K.S. (eds.), *Slope Stability*. Wiley, Chichester, 359–379.
Lehmann, 0. 1933. Morphologische Theorie der Verwitterung von Steinschlagwänden. *Vierteljahresschrift der Naturforschenden Gesellschaft Zürich*, **78**, 83–126.
Linton, D. 1955. The problem of tors. *Geographical Journal*, **121**, 470–487.
Mortensen, H. 1953. Neues zum Problem der Schichtstufenlandschaft — Einige Ergebnisse einer Reise durch den Südwesten der U.S.A., Sommer und Herbst 1952. *Nachrichten d. Akademie der Wissenschaften in Göttingen, Mathematisch-Physikalische Klasse IIa, Jahrgang* 1953, No. 2.
Penck, W. 1924. *Die morphologische Analyse*. Stuttgart, 283 pp.
Phillips, J,.D. and Renwick, W.H. (eds.) 1992. Geomorphic systems. *Geomorphology*, **5**, 195–487.
Pollack, H.N. 1969. A numerical model of the Grand Canyon. In *Fout Corners Geological Society: Geology and Natural History of the Grand Canyon Region*, pp. 61–62.
Rickert, H. 1902. *Die Grenzen der naturwissenschaftlichen Begriffsbildung*, 4th edition. Tübingen.
Römer, W. 1993. Die Morphologie des Alkalikomplexes von Jacupiranga und seiner Umgebung. *Aachener Geographische Arbeiten*, **26**, 271 pp.
Scheidegger, A. 1961. Mathematical models of slope development. *Bulletin Geological Society of America*, **72**, 37–50.
Thomas, M.F. 1974. *Tropical geomorphology*. Macmillans, London, 332 pp.
Thomas, M.F. 1978. The study of inselbergs. *Zeitschrift für Geomorphologie N.F. Suppl. Bd*, **31**, 1–41.
Thorn, C.E. 1988. *Introduction to Theoretical Geomorphology*. Unwin Hyman, Boston, 2347 pp.
Twidale. C.R. 1993. The research frontier and beyond: granitic terrains. *Geomorphology*, **7**, 187–223.
Windelband, W. 1894. *Geschichte und Naturwissenschaft*. Rektoratsrede, Strassburg.
Woldenberg, M. (ed.) 1985. *Models in Geomorphology*. Allen & Unwin, London, 434 pp.
Young, A. 1963. Deductive models of slope evolution. *Nachrichten der Akademie der Wissenschaften in Göttingen, II. Math.-Naturwiss. Klasse*, **5**, 45–66.
Young, A. 1972. *Slopes*. Oliver & Boyd, Edinburgh, 288 pp.

4

Various Aspects of Coral Reef Studies with Special References to Tectonic Geomorphology

YOKO OTA

Department of Geography, Senshu University, Kawasaki, Japan

INTRODUCTION

Coral reef terraces have been studied at many places as useful materials for reconstructing Quaternary sea-level changes in tectonically uplifting areas, because each coral reef terrace not only represents the exact former sea-level position but also provides good age control for the timing of the terrace formation through radiometric dating of the fossil corals. For the same reasons, coral reef terraces also provide the best reference surfaces to detect the pattern and rate of tectonic deformation subsequent to the terrace formation (e.g. Taylor et al. 1980). Therefore, the study of coral reef terraces is very important from the tectonic geomorphological point of view. In this paper, I introduce several aspects of coral reef study related to tectonic geomorphology, taking examples from the western Pacific rim including the Ryukyu and Daito Islands of southwestern Japan, and the Huon Peninsula of Papua New Guinea (PNG) (Figure 4.1). Most of the material used in this paper is from my recent works which were carried out jointly with my colleagues on these areas.

CONTRASTING STYLES AND RATES OF TECTONIC UPLIFT AS DEDUCED FROM CORAL REEF TERRACES ON THE EURASIAN AND PHILIPPINES SEA PLATES, SOUTHWESTERN JAPAN

Among the many islands of the Ryukyu Islands, which are on the overriding Eurasian Plate, west of the plate boundary in the Ryukyu Trench (Figure 4.2), late Quaternary chronologies of coral reef terraces have been established at the three islands of Kikai, Hateruma, and Yonaguni. The chronologies are based on detailed morpho-stratigraphic study and abundant age data by U-series dating (e.g. Konishi et al. 1974; Ota et al. 1982; Omura 1984, 1988). Abundant radiometric ages have been also obtained from two islands, Kita-Daito and Minami-Daito (Figure 4.3) on the subducting Philippine Sea Plate

Geomorphology Sans Frontières. Edited by S. Brian McCann and Derek C. Ford.

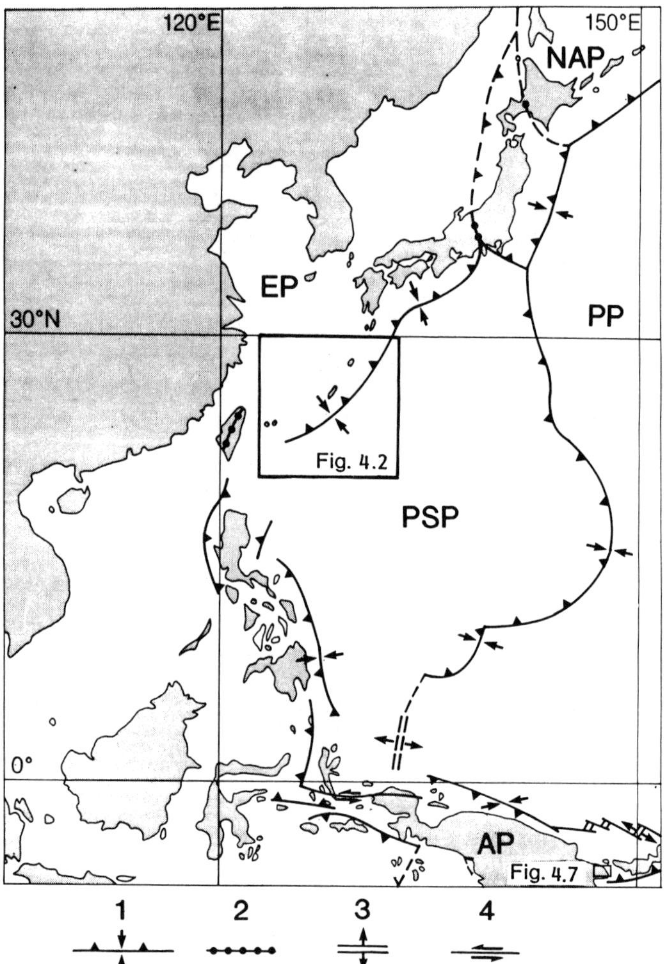

FIGURE 4.1 Map showing the areas discussed in the text and the major plate boundaries on the western Pacific rim. Plate boundaries are after The Research Group for Active Faults of Japan (1991). 1 = subduction zone; 2 = collision zone; 3 = spreading axis; 4 = transform fault; EP = Eurasia Plate; PP = Pacific Plate; NAP = North America Plate; PSP = Philippine Sea Plate; AP = Australia Plate

(e.g. Omura et al. 1991; Ota et al. 1991). There are distinct contrasts in the number of coral reef terraces as well as the pattern and rate of tectonic movements, between islands on the two plates (Figure 4.4). In addition, there are differences among the three islands on the Eurasian Plate, depending on their location relative to the trench axis and the direction of plate motion. Kikai Island of the northern Ryukyus is closest to the Ryukyu Trench and normal to the plate motion. Hateruma of the southern part of the Ryukyu Islands is also close to the trench, but oblique to the plate motion. However, Yonaguni Island is 130 km away from the trench, located instead close to the Okinawa Trough, which is an actively rifting backarc. In the following discussion, the above-mentioned

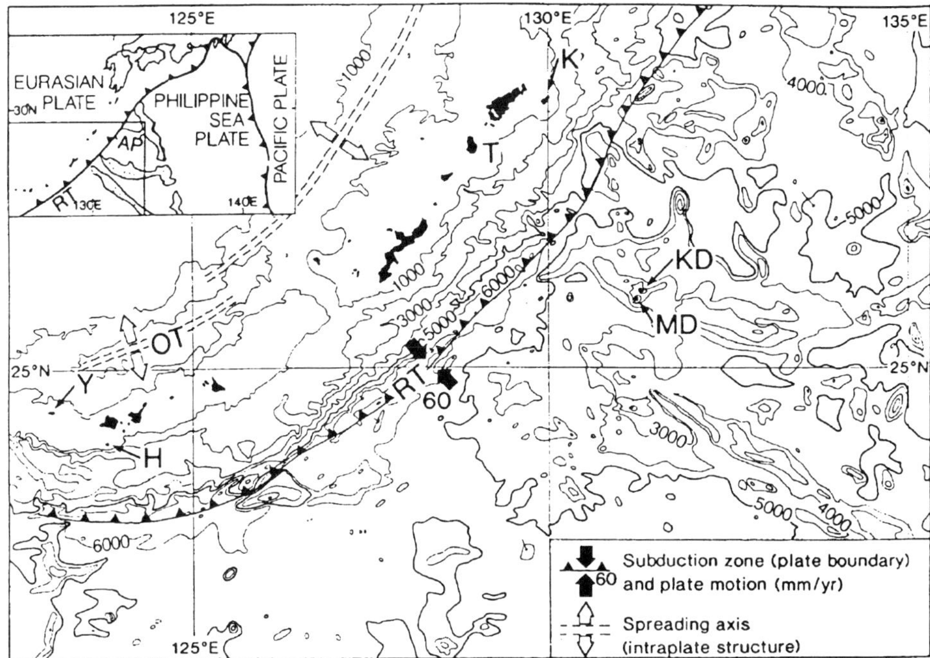

FIGURE 4.2 Tectonic setting and the location of five islands of the Ryukyu and Daito Islands (after Ota and Omura 1992). RT = Ryukyu Trench; OT = Okinawa Trough; K = Kikai Island; H = Hateruma Island; Y = Yonaguni Island; KD = Kita-Daito Island; MD = Minami-Daito Island

five islands are compared in order to understand the different response of each island to the island arc tectonics. This report is mostly based on work by Ota and Omura (1991, 1992).

Comparison of late Quaternary uplift rate and reef terrace sequences

The last interglacial terrace, corresponding to isotope stage 5e of *ca.* 125 ka, is identified on all of these islands as follows (Figure 4.4). The highest terrace of Kikai, which was formed as a table reef and occupies the top of the island, is 224 m above sea level (a.s.l.) (Konishi et al. 1974). The correlative Terrace III of Hateruma Island is the most extensive terrace and ranges in height from 41 m in the east to 22 m a.s.l. in the west (Ota et al. 1982; Omura 1984). The correlative terrace on Yonaguni is the lowest terrace at *ca.* 30 m a.s.l. (Omura et al. 1994). On Kita- and Minami-Daito, very small terraces which are fragmentarily preserved along the coast at only 10–12 m a.s.l. are identified as isotope stage 5e terraces (Kawana et al. 1991; Koba et al. 1991; Omura et al. 1991; Ota et al. 1991).

Thus, the uplift rate since isotope stage 5e has a remarkable range (Table 4.1): 1.8 m/ka for Kikai, 0.3 m/ka for Hateruma, 0.2 m/ka for Yonaguni, and only 0.03–0.05 m/ka for Kita- and Minami-Daito Islands. Obviously the higher uplift rates are those of the islands on the overriding Eurasian Plate, in contrast to the very slow rates of the islands on the subducting Philippine Sea Plate.

FIGURE 4.3 Airphotograph of Kita-Daito Island (Institute of Geographical Survey, Japan). A central depression surrounded by a narrow forested zone is an uplifted atoll, which is fringed by a series of uplifted fringing reefs. The last interglacial terrace of isotope stage 5e is only locally preserved at *ca.* 10 m or lower in altitude on the outer margin of the island

The formation of a flight of terraces is interpreted as the result of tectonic uplift with an appropriate rate at each area superposed on the world-wide eustatic sea-level rise (e.g. Yoshikawa et al. 1964; Bloom 1980; Lajoie 1986). High uplift rates can produce many steps on coasts recording many more of the former sea-level maxima than will be recorded in areas with low uplifting rates. This is apparent in the case of above-mentioned five islands (Figure 4.5). On Kikai, terraces corresponding to isotope stages 5c, 5a, and 3, and even an uplifted Holocene terrace, are present below the highest 5e terrace. In contrast, on Hateruma, only terraces of stages 5c and 5a are developed below the last interglacial terrace. There is no terrace younger than stage 5e on Yonaguni, except for scattered corals with ages corresponding to isotope stage 5c. On Kita- and Minami-Daito, the last interglacial terrace is the lowest of several terraces which have not

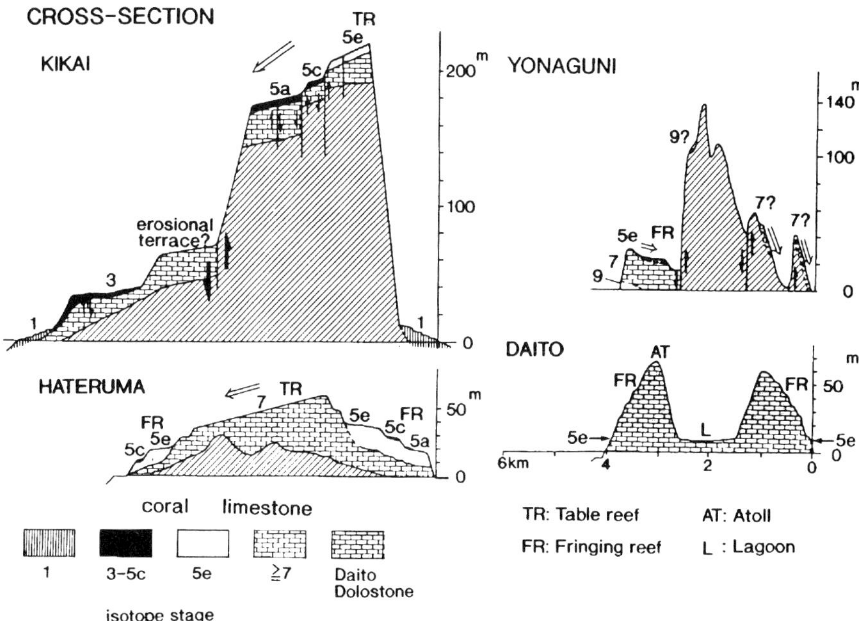

FIGURE 4.4 Schematic geomorphological profiles and geologic sections of four islands on Ryukyu and Daito Islands, southwestern Japan (revised from Omura and Ota 1992 and Ota and Omura 1992)

been dated. It is evident that the uplift rate is an essential controlling factor for the number of preserved terraces on any uplifting coast. This is of course apparent in the case of the even more rapidly uplifting Huon Peninsula coast of Papua New Guinea (e.g. Chappell 1974; Bloom et al. 1974). For example, the southeastern part of the Huon Peninsula, where the uplift rate is 3.4 m/ka (Ota et al. 1993), many terraces corresponding to all of the eustatic high sea levels are preserved as well-defined coral reef terraces (Figure 4.5).

Comparison of patterns of coral reef terraces with uplift history

There is a striking contrast in terrace distribution patterns and uplift history, as summarized in Table 4.1 and Figure 4.4. Kikai Island is characterized not only by high uplift rate, but also by a marked westerly tilt away from the trench side with associated faulting. Thus, the terrace distribution pattern does not show a simple concentric pattern, and terrace correlation is difficult. However, most of the faults have not been active during the Holocene. The emergence of Kikai started only after the formation of a table reef at 125 ka, which now forms the highest terrace. Since then this island has been rising rapidly, with associated active tilting and faulting. Such young and rapid uplift is a characteristic of Kikai, located as it is on the leading edge of the overriding Eurasia Plate.

Hateruma Island is also entirely composed of coral reef terraces. However, the emergence of the island started earlier than Kikai, after isotope stage 7 when the highest

Table 4.1 Comparison of coral reef terraces of five islands of the Eurasian and Philippine Sea Plates (Ota and Omura 1992)

Plate	Overriding Eurasian Plate			Subducting Philippine Sea Plate
Island	Kikai	Hateruma	Yonaguni	Kita- and Minami-Daito
Distance from the trench (km)	80	100	130	150
Age of the oldest reef (ka)	570–840	210	>350	>4,600
Age of oldest reef terrace (ka)	120	210	>350	unknown
Altitude of terrace related to the last interglacial maximum (m)	>224[a]	41[d]	30[c]	10–12
Average uplift rate since the last interglacial maximum (m/ka)	1·8	0·3	0·2	0·03–0·05
Altitude of Holocene terraces (m)	13[b]	—	—	—
Estimated age of emergence of island (ka)	<120	<210	≫350	>460
Evolutional history of reef terraces	table reef→ fringing reef	table reef→ fringing reef	fringing reef	atoll→fringing reef
Style of uplift	rapid uplift with faulting and westerly tilting	slow uplief with small amount of westerly tilting	slow uplift with block movement	very slow uplift

[a] Highest altitude of the table reef terrace.
[b] Highest altitude of the inner margin of fringing reef terrace.
[c] Subdivided into four steps.

terrace was formed. Coral reef terraces on Hateruma show a concentric pattern with slightly longer length in the E–W direction and only slight but progressive westward tilting. Relatively long and slow uplift characterizes this island, despite its close location to the trench.

Coral reef terraces of Yonaguni Island surround the central hills underlain by Tertiary rocks, and were formed as a series of fringing reefs, the oldest of which is probably older than isotope stage 9 (*ca.* 300 ka). The coral reef terraces are deformed into many small tilted blocks bounded by high-angle normal faults. Yonaguni has a longer history as an island and slower uplift. The formation of many distinct tilted blocks on the island reflects the backarc spreading in the adjacent Okinawa Trough.

Both Kita- and Minami-Daito islands (Figure 4.3) have certain common features; they are unique uplifted atolls, each with a central lagoon surrounded by a flight of uplifted fringing reefs. This indicates that the tectonic regime changed from long continued subsidence which resulted in the atoll and lagoon, underlain by a thickness of more than 400 m of coral limestone, to uplift by which a series of fringing reefs has been formed. We have no data to determine when the uplift started. It must be very old, probably older

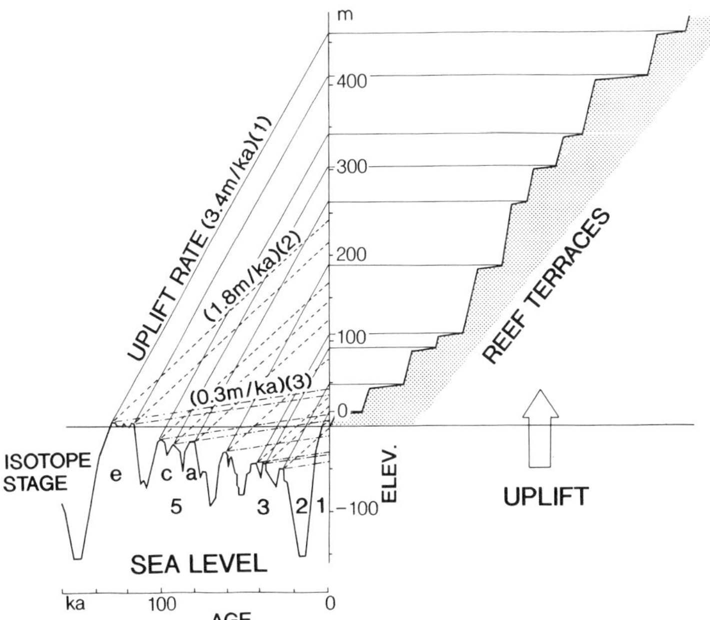

FIGURE 4.5 Diagram showing the superposed result of monotonic land uplift and eustatic sea-level changes for the formation of a flight of coral reef terraces. (1) Bobongara sector (BB) of Huon Peninsula; (2) Kwambu sector (KW) of Huon Peninsula and Kikai Island, northern Ryukyu; (3) Hateruma Island of southern Ryukyu. Sea-level curve is after Bloom et al. (1974)

than 1500 ka, judging by the height of the uplifted atoll (*ca.* 60 m) and the extremely slow uplift rate deduced from the present height of the coral reef of isotope stage 5e.

Interpreting differences in styles and rate of tectonic movement: estimation from coral terrace data

Most of the active deformation of Kikai, characterized by a high uplift rate with tilting and faulting, can be explained by its location on the leading edge of an actively overriding plate. Contrast of the uplift rate between Kikai and Hateruma suggests that differences in the dip angle of the Benioff Zone between the Eurasian and Philippine Sea Plates is responsible, in addition to difference in direction of the plate motion. That is, the steeper dip angle in the southern part of the Ryukyu Trench, near Hateruma, causes easier subduction of the Philippine Sea Plate, than in the northern part, where low-angle subduction near Kikai causes more lateral compression, tilting and rapid uplifting. The slow uplift rate and the formation of tilted blocks on Yonaguni Island, away from the trench, is easily interpreted as a result of backarc spreading.

The change from long continued subsidence to uplift of the Daito Islands could be a result of the formation of a tectonic forebulge associated with northwestward migration of the subducting Philippine Sea Plate toward the Ryukyu Trench. The very low uplift rate since stage 5e suggests that these islands are approaching the time when they begin to subduct beneath the Eurasian Plate. Thus, different tectonic settings of islands control the differential uplift rates and the style of coral reef terrace evolution. In addition, the coral terrace data well constrain otherwise unknown crustal behaviour at plate boundaries.

COSEISMIC UPLIFT OF CORAL REEF TERRACES

Holocene coseismic uplift: Kikai Island and Huon Peninsula, Papua New Guinea

Coseismic uplift associated with historical earthquakes is known at many places in tectonically active areas (e.g. the 1964 Alaskan earthquake, Plafker and Rubin 1978; the 1703 and 1923 earthquakes on the southern Kanto, Matsuda et al. 1978; the 1964 Niigata earthquake, Nakamura et al. 1964, Ota et al. 1988; the 1885 Wairarapa earthquake, Wellman 1967, Moore 1987). It is also known that such coseismic uplift has been repeated during the Holocene and its evidence is preserved as subdivided Holocene terraces (see review by Ota 1991).

Coseismic uplift is well illustrated by coral reef terraces. For example, the Holocene coral reef of Kikai Island is subdivided into four levels (Nakata et al. 1978; Ota et al. 1978; (Figure 4.6); I (9–13 m a.s.l.), II (5– m), III (2.5–5 m) and IV(1.5–2 m), which show a slight westward tilting similar to the Pleistocene terraces. Their ages are 6–6.8 ka for Terrace 1, 3.5–5 ka for II, 3–3.5 ka for III and 1.5–2.5 ka for IV. These steps indicate a repeated episodic uplift of tectonically active Kikai Island during the Holocene, superimposed on a minor eustatic sea-level change, that caused a relatively wide Terrace II with lagoonal deposits and an outer reef edge (Ota et al. 1978). However, there is still a

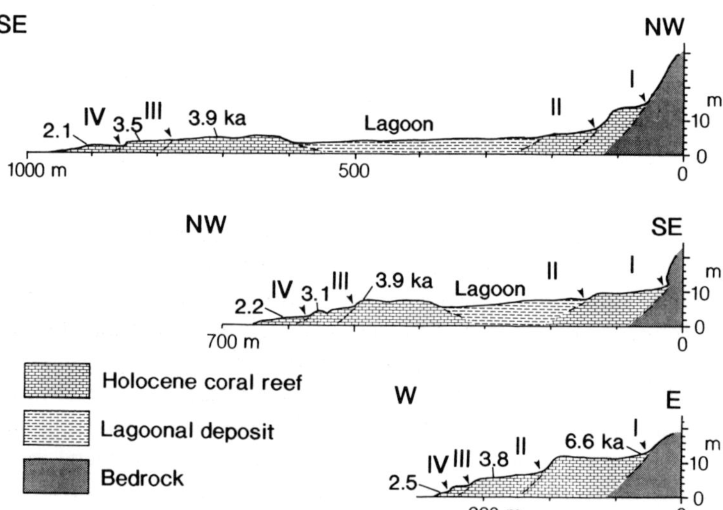

FIGURE 4.6 Examples of cross profile of emerged Holocene coral reefs of Kikai Island and their approximate radiocarbon ages in ka (revised from Ota et al. 1978). SE, NW and W indicate profiles from southeast, northwest and west of the island, respectively

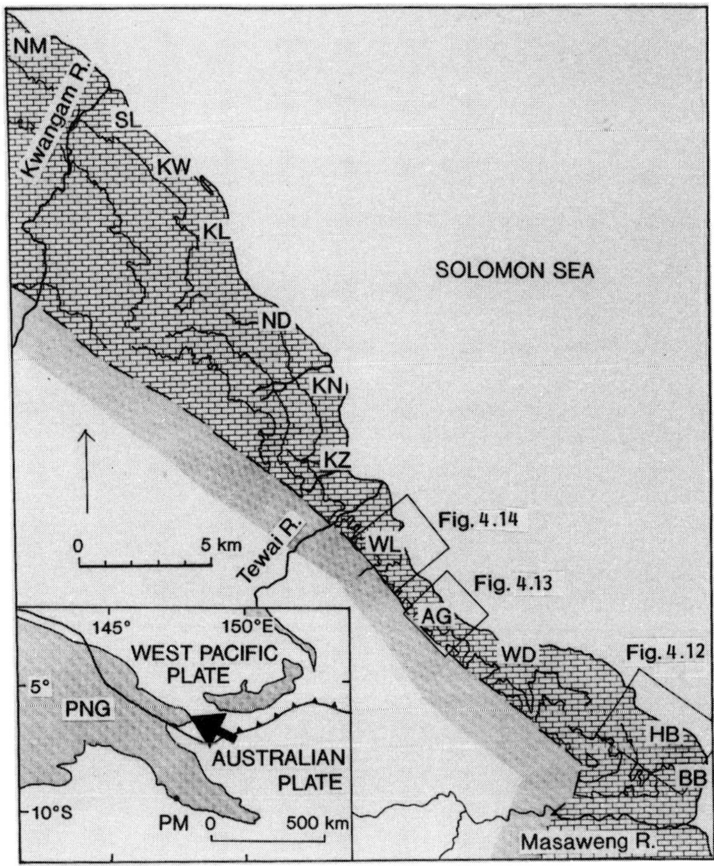

FIGURE 4.7 Index map of the Huon Peninsula, Papua New Guinea with tectonic setting. An arrow in the inset indicates the location of Huon Peninsula. PM = Port Moresby; BB = Bobongara; HB = Hubegong; WD = Wandokai; AG = Agø; WL = Walingai; KZ = Kanzarua; KN = Kanomi; ND = Nanda; KL = Kilasairo; KW = Kwambu; SL = Sialum; NM = Nama

wide age range for each terrace. It is necessary to have more accurate ages for samples obtained from the surface of each terrace, to determine the timing of paleoearthquakes.

Holocene coral reefs of the Huon Peninsula (Figure 4.7), where a late Quaternary chronology of sea-level changes is established on the basis of a detailed study of a flight of coral reef terraces (e.g. Bloom et al. 1974; Chappell 1974; Figure 4.8) are regarded as a key area for this kind of study. They provide a good example of coseismic uplift during the last 6000 years (Ota et al. 1993). The Holocene terrace crest represents the culmination of the postglacial transgression, and ranges in height from 27 m to the southeast to 12 m to the northwest over 40 km of coastline of the Huon Peninsula. A maximum of seven regressive terraces are present below the Holocene crest. Each of the steps is associated with a well-defined reef terrace that is eroded into the Holocene transgressive reef and is accompanied by a notch at its back edge, which represents the former sea-level position (Figure 4.9). The ages of the regressive terraces are known from

FIGURE 4.8 A flight of late Pleistocene coral reef terraces at Sialum of Huon Peninsula. (Photograph by Ota)

FIGURE 4.9 Kwambu coast at Huon Peninsula showing subdivided regressive Holocene terraces. (Photograph by Ota)

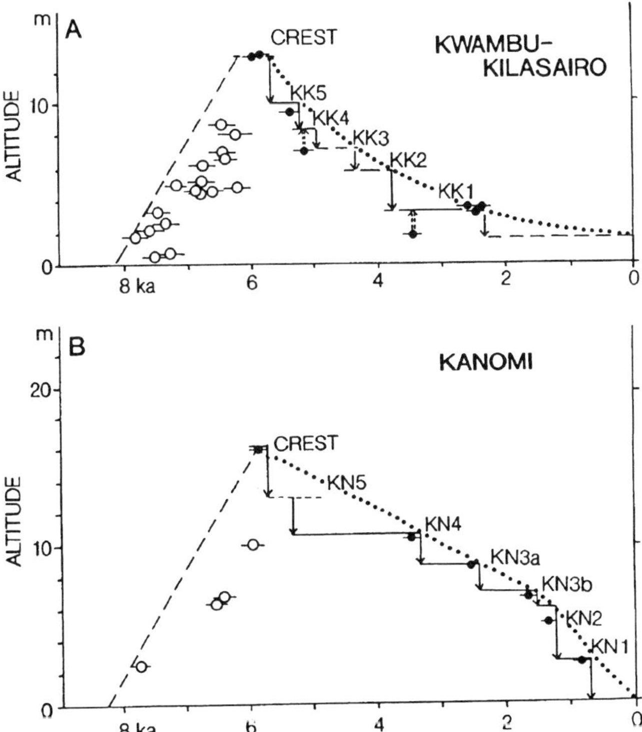

FIGURE 4.10 Age and height plots of Holocene coral reef terraces at two sites from the Huon Peninsula (after Chappell et al. in press). Samples older than *ca.* 6 ka (open circles) represent Holocene transgression, and those younger than *ca.* 6 ka (solid circles) were obtained from regressive corals associated with episodic coseismic uplift

^{14}C ages of attached corals that grew on the terraces and are significantly different from the ages of the adjoining terraces (Figure 4.10). Thus, it is evident that the high uplift rate during the middle Holocene on the Huon Peninsula was characterized by a series of episodic uplifts, probably associated with earthquakes (Ota et al. 1993).

The age of the lowest terrace varies from place to place as follows: *ca.* 2.5 ka in the northwest (KK), 0.8 ka in the central region (KN) and the southeast (WD) (Figure 4.10). The range in ages indicates that this coastal zone is segmented into at least three subregions that have different uplift history (Figure 4.10), in a fashion similar to that described by Berryman et al. (1989) for the Holocene terrace sequence along the eastern coast of North Island, New Zealand. The exact subregion boundaries and the location of seismogenetic faults are not known, however.

Late Quaternary coseismic uplift identified on the Huon Peninsula

The northeast coast of the Huon Peninsula, where a flight of coral reef terraces extends up to 1000 m in altitude and dates back to middle Quaternary is well preserved (Figure 4.8)

and is well known as a rapidly uplifting area with a maximum uplift rate of 3.5 m/ka. Ota et al. (1993) indicated that the long-term uplift rate since the formation of reef complex VII (*ca.* 125 ka, isotope stage 5e) was the same as the uplift rate since the middle Holocene (reef complex I, *ca.* 6 ka) in the northeastern Huon Peninsula (Table 4.2). This suggests that the same tectonic processes and deformation pattern have continued in that area throughout the late Quaternary. Therefore, coseismic uplift similar to that detected from the Holocene terrace should be present in the late Pleistocene terraces, as already mentioned by Chappell (1983).

Profiles of late Pleistocene terraces were surveyed at three locations, which have different uplift rates, using an Electric Distance Meter, in order to identify expected coseismic terraces within the late Pleistocene terraces (Ota and Chappell 1996). Figure 4.11 shows a profile of the terraces lower than IIIa. There are many small steps in the profile, in addition to the obvious major terrace risers between the coral reef terraces which were formed during major late Pleistocene transgression (IIIa, IIIb, II and I; Chappell et al. 1995). These many small steps cannot be interpreted as the result of eustatic sea-level changes. The spacing between successive small terraces ranges from 3 to 6 m, similar to the separation of the Holocene terraces. Therefore, it is possible to conclude that each of these small steps is the result of a single coseismic uplift event. Recurrence intervals of major earthquakes during terrace IIIa and II are estimated to be 1250–1450, 1000–1900, and 1400–2800 years for Bobongara, Tewai and Kanzarua, judged from the possible number of small terraces (Ota and Chappell, 1996). These intervals are little different from intervals deduced from Holocene terraces, considering that the Pleistocene estimates are likely to be raised by erosional loss of small terraces.

Repeated coseismic uplift during the late Quaternary is known from many other places, such as Sado Island and Awashima Island in the Sea of Japan, off northern Japan (Ota et al. 1976, 1988); Muroto Peninsula, situated on the Pacific side of southwestern Japan (Yoshikawa et al. 1964); and the east coast of North Island, New Zealand (Yoshikawa et al. 1980). In each of these areas, progressive deformation has a similar pattern to that of

TABLE 4.2 Comparison of mean Holocene and late Pleistocene uplift rates of Huon Peninsula (Ota et al. 1993)

Site	Altitude of Holocene crest (m)	Mean Holocene uplift rate (m/k)a[a]	Altitude of Pleistocene crest (m)	Mean Pleistcene uplift rate (m/ka)[b]
Kwambu	13·2	1·9–2·0	227	1·8–1·9
Blucher	15·5	2·2–2·3	280	2·2–2·4
Kanzarua	20·0	2·8–3·0	340	2.7–2.9
Hubegong/ Bobangara[c]	23·0	3·2–3·4	406	3·2–3·4

[a] Holocene uplift rates based on calibrated age of crest = 5700–7100 years equivalent to ^{14}C age of 6000–6400 years BP.
[b] Pleistocene uplift rates based on age of Reef VII crest = 118 000–126 000 years, for surveyed benchmarks at locations shown in Figure 4.7.
[c] Hubegong for the Holocene and Bobabgara nearby for the last interglacial terrace.

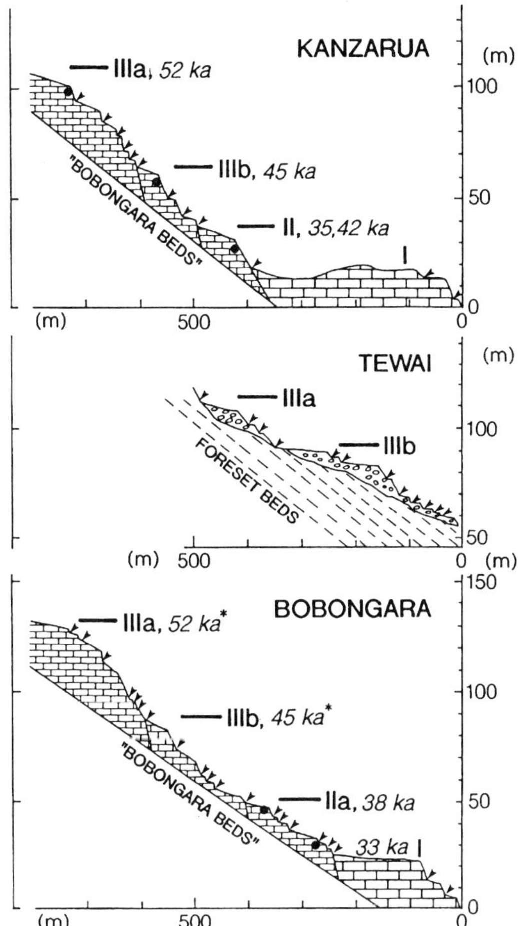

FIGURE 4.11 Profiles of late Quaternary coral reef terraces younger than Terrace IIIa (*ca.* 52 ka) at Bobongara, Tewai and Kanzarua of Huon Peninsula. Arrows indicate the inner margin of all the terraces identified in the field. Dashed lines represent major boundaries of transgressive coral limestones of eustatic origin. Most of terraces are regressive and coseismic origin. (After Ota and Chappell 1996)

historical earthquakes or Holocene terraces. However, it has been difficult to identify each small terrace that represents a single episodic uplift, because each terrace, which records a rather short stable relative sea level, was smoothed by secondary cover deposits such as loess or tephra, or was rapidly eroded, and small steps of the terrace surface have been smoothed by subsequent weathering and erosion in these areas where the terraces are underlain by Tertiary sedimentary rocks and capped only by thin unconsolidated beach deposits.

The present height of terraces can be interpreted as the cumulative result of repeated fluctuation of eustatic sea level and monotonic tectonic uplift, the rate of which naturally varies from place to place, as already mentioned (Figure 4.5). However, it must be noted that monotonic constant uplift cannot be represented as a straight line, but as a succession of stepwise uplifts, each separated by a short time span.

LANDSLIDES AS A POSSIBLE INDICATOR OF PALEOEARTHQUAKES; A CASE STUDY FROM THE HUON PENINSULA

Many landslides occur on the coral reef terraces of the Huon Peninsula. The shape and pattern of the landslides, as well as their density and age of their occurrence, show great local variation (Ota et al. 1995; Figures 4.12, 4.13, 4.14). We tentatively classify them into two types, based primarily on the presence or absence of debris-flow deposits associated with each landslide, and secondarily on their extent, depth and other geomorphic aspects (Figure 4.15). Type I is a landslide that has a well-identified theatre-like head scar without significant debris flow. This type can be subdivided into the following four subtypes: Type la has only collapsed blocks of Pleistocene limestone (e.g. O and P of Figure 4.13; R of Figure 4.14; Figure 4.16). Type lb is similar to la, but is accompanied by small amount of debris flow which never flooded out through gullies into lower stream channels. Type lc is a landslide that has large rotated collapse blocks including bedrocks beneath the Pleistocene coral limestone. Occasionally younger terrace sequences have developed on such huge blocks (Q of Figure 4.14). Type Id is a shallow landslide with slumped blocks on which the original terrace sequences are still preserved (S of Figure 4.14). Type II is a landslide that produces a considerable amount of debris flow accompanied by fluvial debris outflow, resulting in a levee-like bulge or lobe on the coral reef terrace (e.g. A and B of Figure 4.12; M and N of Figure 4.13; Figure 4.17). Landslides of the above-mentioned types formed in late Quaternary times during the formation of the coral reef terraces and most of them are now stabilized. There are also many other landslides, which are not discussed in this paper and are found on the steep scarps along valley walls or at steep valley heads, including modern landslides. They are formed in association with gully incision and headward erosion.

Landslides can be caused by heavy rainfall or oversteepening of land related to active headward- or gully erosion, but large earthquakes also can trigger large-scale landslides, such as debris flow that occurred during the May 1992 earthquake and those landslides that resulted from the 1985 earthquake. Paleolandslides which are recorded on landforms are considered to be one of the major tools for reconstructing large past earthquakes (e.g. Crozier 1991).

We particularly examined the ages of landslides on the coral reef terraces of the Huon Peninsula, in order to estimate the age of paleoearthquakes, by ^{14}C dating of corals above or below debris-flow deposits. Ages of some landslides can be also estimated from morpho-stratigraphical relationships between landslide features and dated coral terraces. The timing of 27 major landslides examined in the seven sectors of the Huon Peninsula is summarized in Table 4.3, and is classified into four age groups (Ota et al. 1995)

— *Group a*: younger than *ca.* 1500 yr BP. The youngest group of landslides identified in the Huon Peninsula is represented by a debris flow of the Ago River mouth (M of Figure 4.13). This debris flow immediately rests on beach deposits dated at *ca.* 1250 yr BP, which can be correlated with the youngest Holocene regressive terrace nearby.
— *Group b*: 6500–9000 yr BP. The next younger group, which is identified from ^{14}C

ages of corals, interfingered with debris flow deposits (e.g. Loc. 1 and 2 of Figure 4.12; Figure 4.18). This group occurred during the late stage of the Holocene marine transgression.

— *Group c*: 7 to 33 ka. Many landslides disrupt terrace II (*ca*. 33 ka) and older terraces, and are trimmed by the Holocene shoreline that represents the postglacial transgression (e.g. O and P of Figure 4.13).

— *Group d*: older than 33 ka. There are also many landslides which occurred prior to the formation of terrace II.

At this stage, it is impossible to say that all of landslides in the Huon Peninsula were triggered by large earthquakes, because the timing of landslides belonging to Groups c and d is not specified, and can be subdivided further into several age groups when better

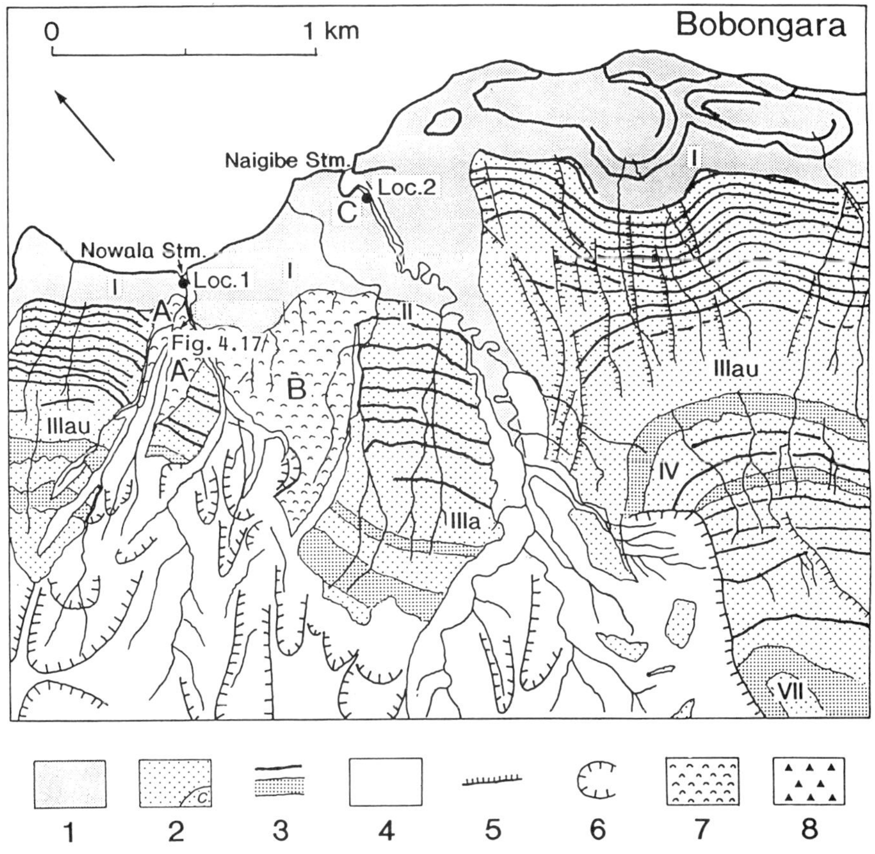

FIGURE 4.12 Geomorphological map of the Bobongara area showing terrace sequence and landslides (Ota et al. 1995). 1 = Holocene coral terraces; 2 = Pleistocene coral terraces (c, crest); 3 = terrace risers; 4 = slopes and ridges; 5 = faults; 6 = slumping scarps of landslides; 7 = debris flows; 8 = slumped blocks. These symbols are common for Figures 4.13 and 4.14

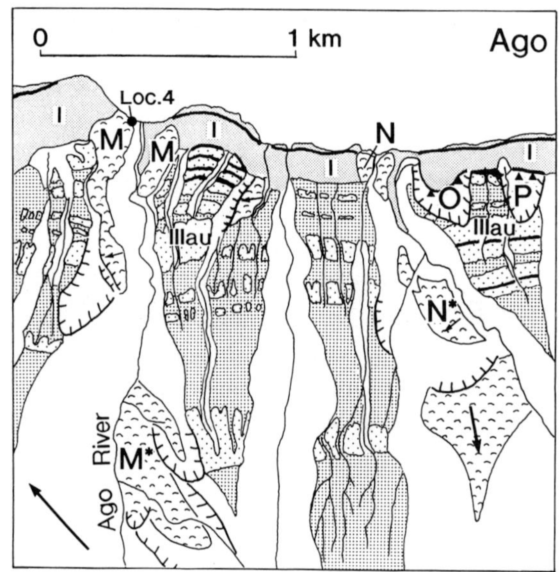

FIGURE 4.13 Geomorphological map of the Ago area showing terrace sequence and landslides (Ota et al 1995)

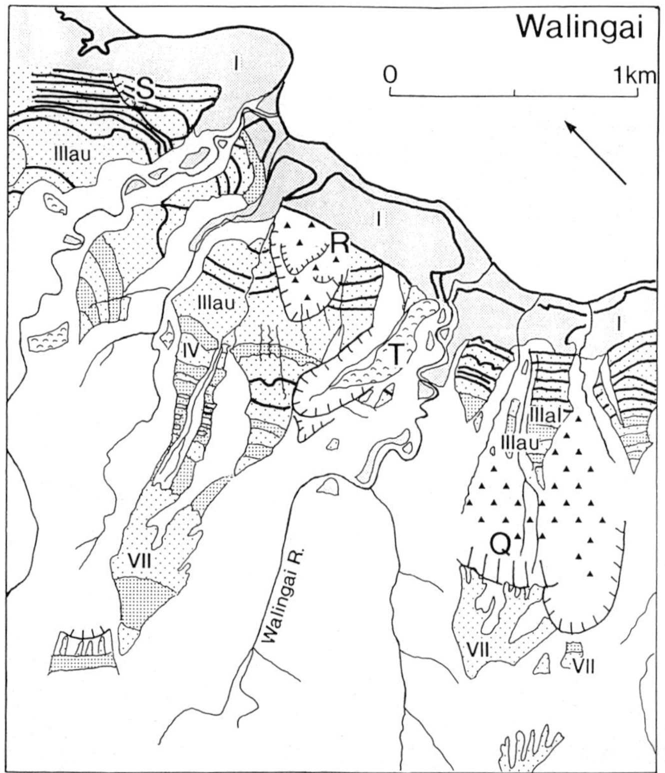

FIGURE 4.14 Geomorphological map of the Walingai area showing terrace sequence and landslides (Ota et al. 1995)

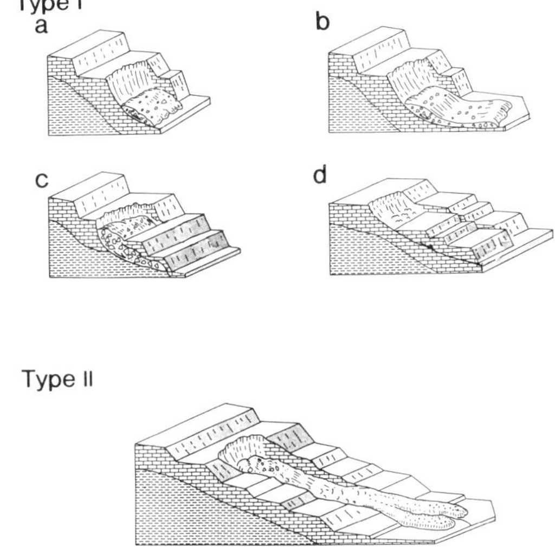

FIGURE 4.15 Schematic diagrams showing various types of landslides on the Huon Peninsula (Ota et al. 1995)

FIGURE 4.16 Landslide of Type Ia at Walingai (R of Figure 4.14), belonging to age group c. (Photograph by Ota)

FIGURE 4.17 Debris-flow levee on coral reef terraces on the Bobongara area at Huon Peninsula (A of Figure 4.12). (Photograph by Ota)

age control is given for each of them. However, coseismic origin for landslides of Groups a and b is strongly considered on the basis of that (1) the timing of Group a is identical to the age of the emergence of the youngest regressive coseismic terrace on the southeastern part of the Huon Peninsula (Chappell et al. in press), and (2) the large debris flows which are interfingered with Holocene transgressive corals and have nearly the same ^{14}C ages are identified at several locations (Ota et al. 1995).

Density of landslide distribution has a significant difference on the northwest and southeast of Kanomi (KN in Figure 4.7). There are many landslides on the southeastern area with higher uplift rate (more than 2.5 m/ka) than on the northwestern area with relatively lower uplift rate. Particularly, distribution of large-scale landslide of Type II is confined to the southeastern area (Ota et al. 1995). This implies that the higher uplift rate of the southeastern area provides a suitable situation for the landslide occurrence through

TABLE 4.3 Age group of landslides on the coral reef terraces of Huon Peninsula (simplified from Ota et al. 1995)

Group by age	Estimated age	Number
a	$\leq 1 \cdot 5$ ka BP	3
b	*ca.* 6·5 ~ 8 ka BP	5
c	post II, pre I (*ca.* 33~ 6 ka BP)	10
d	pre II ($\geq$ 33 ka)	9

FIGURE 4.18 Debris flow deposits on regressive beach deposits of *ca.* 1250 yr BP on the Ago River mouth, Huon Peninsula (M of Figure 4.13). (Photograph by Ota)

its more frequent seismicities, and the formation of steeper slopes. Therefore, tectonic control is a still very important factor for the formation of large landslides, even though their coseismic origin is not always confirmed.

CONCLUDING REMARKS

I have reviewed some aspects of coral reef studies related to tectonic geomorphological approach, including some of our recent works. Major topics are summarized as follows:

(1) Contrasting styles and rates of tectonic deformation of coral reef terraces is presented from three islands of the Ryukyus on the overriding Eurasia Plate and two islands of the Daito Islands on the subducting Philippine Sea Plate. The results constrain some of the unknown nature of the plate boundary.

(2) Coseismic origin for the formation of many small regressive terraces is identified at the coasts with high uplift rate. Late Pleistocene coseismic uplift is proved on the Huon Peninsula, giving nearly the same amount and recurrence interval of coseismic uplift throughout the late Quaternary. Coseismic uplift is regarded as an important tectonic process in rapidly uplifting coast.
(3) Some of paleolandslides on the coral reef terraces can be used as an indicator for reconstructing the timing of paleolearthquakes. Distribution and density of large paleolandslides also reflect the regional difference of tectonic uplift.

However, further detailed study on morpho-stratigraphical observations of coral terraces and accurate radiometric dating for all the terraces are required in order to establish the Quaternary chronology, which provides the most fundamental basis for a tectonic geomorphological approach. Also more systematic study of a wide area, including many coral reef terraces which have a high potential for this kind of study, but were not examined in detail, is required for generalization of several points discussed in this paper.

ACKNOWLEDGEMENTS

I am grateful to many colleagues for joint works in the field and for valuable discussion during the course of this study. Special thanks are due to John Chappell for recent joint works in the Huon Peninsula and Akio Omura for long years of joint study in the Ryukyu and Daito Islands. I would like to express my thanks to Olav Slaymaker who invited me to give this paper as one of the plenary lectures of the Third International Geomorphology Conference. I am indebted to Art Bloom who kindly corrected and reviewed the first draft and Takatoshi Fujimori and Kumiko Ikura for drawing some of the figures.

REFERENCES

Berryman, K.R., Ota, Y. and Hull, A.G. 1989. Holocene paleoseismicity in the fold and thrust belt at the Hikurangi subduction zone, eastern North Island, New Zealand. *Tectonophysics*, **163**, 185–195.

Bloom, A.L. 1980. Late Quaternary sea level changes on south coasts: A study in tectonic diversity. In Morner, N.A. (ed.), *Earth Rheology, Isostasy and Eustasy*, Wiley, Chichester, pp. 505–516.

Bloom, A.L., Broecker, W.S., Chappell, J., Matthews, R.K. and Mesollela, K.J. 1974. Quaternary sea level fluctuations on a tectonic coast: new ^{230}Th/^{234}U dates from the Huon Peninsula, New Guinea. *Quaternary Research*, **4**, 185–205.

Chappell, J. 1974. Geology of coral terraces at Huon Peninsula, New Guinea: a study of Quaternary tectonic movements and sea level changes. *Geological Society of America Bulletin*, **85**, 553-570.

Chappell, J. 1983. A revised sea-level record for the last 300,000 years from Papua New Guinea. *Search*, **14**, 99–101.

Chappell, J. and Shackleton, N. J. 1986. Oxygen isotopes and sea level. *Nature*, **324**, 137–140.

Chappell, J., Ota, Y. and Berryman, K.R. Late Quaternary coseismic uplift history of Huon Peninsula, Papua New Guinea. *Quat. Sci. Rev.* (in press).

Chappell, J., Omura, A., MuCulloch, M., Esat, T., Pandolfi, J., Ota, Y. and Pillans, B. 1995. Reconciliation of late Quaternary sea levels derived from coral terrace at Huon Peninsula with deep sea oxygen records. *Journal of Geography*, **104**, 777–784 (in Japanese with English abstract).

Crozier, M. 1991. Determination of paleoseismicity from landslides. In Bell, D.H. (ed.), *Landslides*. Balkema, pp. 1173–1180.

Kawana, T., Taira, H., Tanahara, A., Aoki, H., Ota, Y., Omura, A. and Koba, M. 1991. ^{226}Ra/^{238}U dates by non-destructive-ray spectrometry of late Pleistocene coral in Minami- and Kita-Daito Island, Okinawa, Japan. *Journal of Geography*, **100**, 367–377 (in Japanese with English abstract).

Koba, M., Tamura, M., Kaigara, T., Ikeda, S., Ota, Y. and Omura, A. 1991. Electron spin resonance coral ages obtained from a raised atoll, Kita-Daito Island, in the north Philippine Sea. *Journal of Geography*, **100**, 351–366 (in Japanese with English abstract).

Konishi, K., Omura, A. and Nakamichi, O. 1974. Radiometric coral age and sea level records from the late Quarternary reef complexes of the Ryukyu Islands. *Proceedings of the 2nd International Coral Reef Symposium*, **2**, 595–613.

Lajoie, K.L. 1986. Coastal tectonics. In Wallace, R. (ed.), *Active Tectonics*. Academic Press, pp. 95–124.

Matsuda, T., Ota, Y., Ando, M. and Yonekura, N. 1978. Fault mechanism and recurrence time of major earthquakes in southern Kanto district, Japan, as deduced from coastal terrace data. *Geological Society of America Bulletin*, **89**, 1610–1618.

Moore, P. R. 1987. Ages of the raised beaches of Turakirae Head, New Zealand. *Journal of the Royal Society of New Zealand*, **17**, 313–324.

Nakamura, K., Kasahara, K. and Matsuda, T. 1964. Tilting and uplift of an island, Awashima near the epicenter of the Niigata Earthquake in 1964. *Journal of Geodetical Society of Japan*, **10**, 172–179.

Nakata, T., Takahashi, T. and Koba, M. 1978. Holocenc emerged coral reefs and sea-level changes in the Ryukyu Islands. *Geographical Review of Japan*, **51**, 87–108 (in Japanese with English abstract).

Omura, A. 1984. Uranium-series age of the Riukiu Limestone on Hateruma Island, southwestern Ryukyus. *Transactions and Proceedings of the Paleontological Society of Japan, New Series*, **135**, 415–426.

Omura, A. 1988. Geologic history of the Kikai Island, central Ryukyus, Japan: summary of uranium series dating of fossil corals from the Riukiu Limestone. *The Memoirs of the Geological Society of Japan*, **29**, 253–268 (in Japanese with English abstract).

Omura, A. and Ota, Y. 1992. Paleo sea level changes during the last 300,000 years deduced from the morpho-stratigraphy of coral reef terraces and $^{230}Th/^{234}U$ age of terrace deposits. *Quaternary Research*, **31**, 313–327 (in Japanese with English abstract).

Omura, A., Iwata, H., Ota, Y., Koba, M. and Kawana, T. 1991. $^{230}Th/^{234}U$ dates of late Pleistocene corals from Kita- and Minami-Daito Islands, Okinawa, Japan. *Journal of Geography*, **100**, 337–350 (in Japanese with English abstract).

Omura, A., Kodama, K., Watanabe, M., Suzuki, A. and Ota, Y. 1994. Tectonic history of Yonaguni Island, Southwestern Ryukyus, Japan, deduced from coral reef terraces and uranium series dates of Pleistocene corals. *The Quat. Res. Japan*, **33**, 213–231 (in Japanese with English abstract).

Omura, A., Chappel, J., Pillans, B., MuCulloch, M., Esat, T., Sasaki, K. and Kawada, Y. 1994. Alpha-spectrometric $^{230}Th/^{234}U$ dating of Pleistocene corals from Huon Peninsula. In Ota, Y. (ed.), *Preliminary Reports on the Study of Coral Reef Terraces of Papua New Guinea*, Yokohama National University, pp. 95–110.

Ota 1991. Coseismic uplift in coastal zone of the western Pacific rim and its implications for coastal evolution. *Zeitschrift für Geomorphologie N.F., Suppl.-Bd*, **91**, 163–179.

Ota, Y., Matsuda, T. and Naganuma, K. 1976. Tilted marine terraces at the Ogi Peninsula, Sado Island, central Japan, related to the Ogi earthquake of 1802. *Jisin, Ser.* 2, **29**, 55–70 (in Japanese with English abstract).

Ota, Y., Machida, H., Hori, N., Konishi, K. and Omura, A. 1978. Holocene raised coral reefs of Kikai-Jima (Ryukyu Islands). *Geographical Review of Japan*, **51**, 109–130 (in Japanese with English abstract).

Ota, Y., Hori, N. and Omura, A. 1982. Age and deformation of marine terraces of Hateruma Island, southwestern Japan. *Abstracts of XI INQUA Congress*, Moscow, **2**, 232.

Ota, Y. and Omura, A. 1991. Uplift rate comparison based on coral reef terraces on different plates, southwestern Japan. *Bulletin of the INQUA Neotectonics Commission*, **14**, 13–15.

Ota, Y. and Omura, A. 1992. Contrasting styles and rates of tectonic uplift of coral reef terraces in the Ryukyu and Daito Islands, southwestern Japan. *Quaternary International*, **15/16**, 17–29.

Ota, Y., Kashiwagi, S., Sakurai, I. and Ikeda, J. 1988. Coseismic deformation of late Quaternary marine terraces at Awashima Island, off Niigata coast, central Japan. *Journal of Geography*, **97**, 25–38 (in Japanese with English abstract).

Ota, Y., Omura, A., Koba, M., Kawana, T. and Miyauchi, T. 1991. Late Quaternary tectonic movement as deduced from raised coral reefs of Daito islands on northwestern part of the Philippine Sea plate. *Journal of Geography*, **100**, 317–336 (in Japanese with English abstract).

Ota, Y., Chappell, J., Kelley, R., Yonekura, N., Matsumoto, E., Nishimura, T. and Head, J. 1993. Holocene coral reef terraces and coseismic uplift of Huon Peninsula, Papua New Guinea. *Quarternary Research*, **40**, 177–188.

Ota, Y., Chappell, J., Berryman, K.R. and Okamoto, Y. 1995. Distribution and genesis of large-scale landslides disrupting coral terraces at Huon Peninsula, Papua New Guinea. *Journal of Geography*, **104**, 684–705 (in Japanese with English abstract).

Ota, Y. and Chappell, J. 1996. Late Quaternary coseismic uplift events on the Huon Peninsula, Papua New Guinea, deduced from coral terrace data. *Journal of Geophysical Research*, **10/B3**, 6071–6082.

Plafker, G. and Rubin, M. 1978. Uplift history and earthquakes deduced from marine terraces on Middleton Island, Alaska. *USGS Open-file Report*, **79-943**, 678–721.

Research Group for Active Faults of Japan 1991. *Active Faults in Japan – sheet maps and inventories*, University of Tokyo Press.

Taylor, F.W., Isacks, B.L., Jounannic, C., Bloom, A.L. and Dubois, J. 1980. Coseismic and Quaternary vertical movements, Santo and Malekula Islands, New Hebrides Island Arc. *Journal of Geophysical Research*, **85-B10**, 5367–5381.

Wellman, H.W. 1967. Tilted beach ridges at Cape Turaakirae, New Zealand. *Journal of Geosciences, Osaka City University*, **10**, 123–129.

Yoshikawa, T., Kaizuka, S. and Ota, Y. 1964. Mode of crustal movement in the late Quaternary on the southwest coast of Shikoku, southwestern Japan. *Geographical Review of Japan*, **37**, 1–22 (in Japanese with English abstract).

Yoshikawa, T., Ota, Y., Yonekura, N., Okada, A. and Iso, N. 1980. Marine terraces and their deformation on the northeast coast of the North Island, New Zealand. *Geographical Review of Japan*, **53**, 238–262 (in Japanese with English abstract).

5

Impact of Mountain Glaciations on Tors, Blockfields and Cryoplanation Features. Nunataks or Non-scoured Zones as Refugia?

ANDERS RAPP

Department of Physical Geography, University of Lund, Sweden

INTRODUCTION

An unexpected event of great importance to stimulate research on mountain geomorphology is the recent discovery of a frozen man from the stone age at the bottom of a glacier in the European Alps. I refer to the so-called Iceman Ötzi, who was found on the 19 September 1991 near the frontier between Austria and Italy at 3200 m altitude and who was later dated by radiocarbon dating. He died about 5300 years ago. If that find is investigated, interpreted and reported on properly it can bring new light on many questions related to mountain research: geomorphology, climatology, biology and environmental history. This find can, for instance, help to revive and increase general interest in the scientifically unsolved problem of how life and landforms survived periods of glaciation in mountains.

When I was a student at Uppsala university in the 1950s, the so-called "nunatak problem" of supposed ice-free refugia in Scandinavian mountains was one of the leading themes in mountain geomorphology and in landscape ecology. The nunatak problem has been a scientific subject of much controversy for about 100 years now. In the 1950s, two of the most famous defenders of the concept of persistent nunataks as refugia for plants and animals were the botanists Rolf Nordhagen and Eilif Dahl, professors in Norway. I heard both of them as guest lecturers in Uppsala and recall Nordhagen's concluding remark in a discussion in the mid-1950s: "I have my firm conviction that certain plants did survive the main glaciations on ice-free refugia. And I will die happy in that belief". I tell you this memory as an example of the strong feelings, deep interest and interdisciplinary debate involved in the 100-year controversy on how life survived glaciations.

Many geo-scientists believed in the counter-argument, that long-distance transported glacial erratics which were found high up on the postulated nunataks and refugia were proof of an earlier ice cover in one or several glacial maxima and thus extinction of all

Geomorphology Sans Frontières. Edited by S. Brian McCann and Derek C. Ford.

life — the so-called "tabula rasa" theory. As G. Hoppe expressed it in a paper in the Swedish NFR yearbook of 1959:

> One of the strongest arguments for the refugia hypothesis is the existence of plants and animals with a bicentric distribution. There may however be other ways of explaining such distributions. Attention is called to the fact that the supposed refugia could have been deglaciated at a much earlier stage than the rest of the country and thus offered both flora and fauna a longer time to get established.
>
> (Hoppe 1959, with reference to a series of articles by biologists and geo-scientists in 1958 and 1959.)

In America a thorough and well-known review on the nunatak controversy was published by J.D. Ives in 1974. He finished his article with the following words: "... I am deeply grateful for fifteen years of encouragement, stimulation and criticism from a small group of the major contributors to the nunatak discussion: to Professors Eilif Dahl, Gunnar Hoppe, Carl H. Lindroth, Askell Löve and Drs. Ragnar Dahl and Doris Löve, protagonist and antagonist alike." Twenty years later this truly international and transdisciplinary discussion still continues, e.g. with stimulating inputs in articles by the late Eilif Dahl in the 1980s and a review by K. Faegri in 1993.

My approach to the nunatak problem has mainly been to try to learn and make reasonable interpretations from geomorphological evidence of geomorphic processes and landforms at high levels. Initially I regarded the alpine blockfields as the zone of maximum postglacial frost weathering. Now I think, on the contrary, they can be a zone of both very weak glacial scour during the glaciations due to protecting cold-based ice on permafrost and also very slow periglacial or postglacial activity. Permafrost conditions, few freeze–thaw cycles and short snow-free period are parts of the explanation. Permafrost can perhaps preserve seeds of plants for thousands of years in sheltered pockets beneath cold-based ice or in morainic debris on the surface of glacier ice (cf. Porsild et al. 1967).

Figure 5.1 shows a northern unicentric species *Cassiope tetragona*, which has scattered occurrences as far east as the Kola peninsula (Hultén 1950). On both sides of the Atlantic a typical amphiatlantic species is *Salix herbacea* "that probably survived the Weichsel–Wisconsin maximum in the North Atlantic area" to quote the review by Ives (1974).

GEOMORPHOLOGICAL IMPACT OF COLD-BASED GLACIERS ON PERMAFROST

In the 1960s I and several of my Scandinavian colleagues in geomorphology were of the opinion that most of the high-alpine blockfields in Scandinavia were formed by vigorous frost-shattering in postglacial time, after being completely covered by glaciers in the Pleistocene glaciations — and also by deposition of till and erratics. But there were many unsolved problems and open questions with this "tabula rasa" theory and also with the nunatak theory on geomorphological grounds, so we continued to search for improvements in our interpretations. I remarked in my doctor's thesis (Rapp 1960) that three indicators of very slow processes in blockfields were old lichen cover, lakes in the blockfield zone with very slow sedimentation, and fresh-looking glacial drainage channels — formed *ca.* 9000 BP — in some blockfields. I also found that rockwall

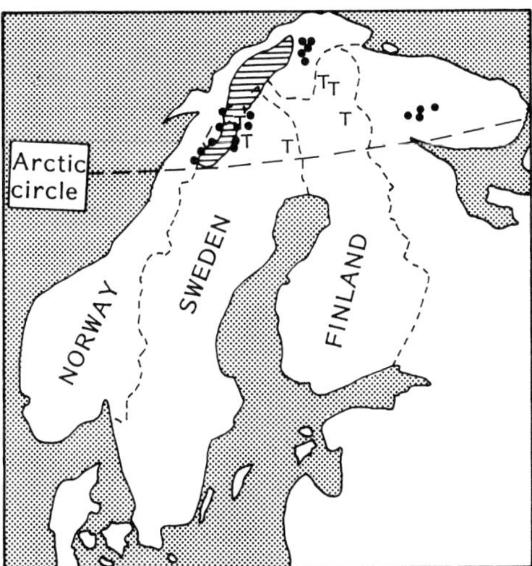

FIGURE 5.1 Unicentric distribution of a mountain plant — *Cassiope tetragona* (White arctic bell-heather) — in northern Scandinavia (after Hultén 1950). Note the eastern scattered occurences (dots), possibly near glacial refugia or non-scoured zones. T-symbols show pre-glacial in tors in northern Finland (Seppälä 1986) and in northeastern Sweden (Kleman, oral information, 1995)

funnels and chutes were mainly eroded before the latest deglaciation and probably had survived the glaciation due to weak glacial scour (Rapp 1960, Spitsbergen and Pallenvagge, Abisko). Ragnar Dahl introduced considerable improvements with his theory on "Shifting ice culminations, alternating ice cover and ambulant refugia organisms" (Dahl 1963). Sugden and Watts (1977) made a very interesting contribution with their interpretation of the genesis of blockfields on Baffin Island in Arctic Canada. I quote "On the basis of two case studies in eastern Baffin Island it is suggested that tors and felsenmeer have survived beneath the Laurentide ice sheet. These features therefore, need not be diagnostic of areas that escaped glaciation. Rather, the existence of tors and felsenmeer may reflect situations where the overlying ice sheet was cold-based?"

The introduction of the concept of cold-based ice and the growing and improved knowledge of the distribution of mountain permafrost today in Scandinavia (King 1986) requires re-consideration of the nunatak theory and alpine blockfield genesis (Figure 5.2).

I have drawn attention to this ongoing change of paradigm in mountain geomorphology in a recent paper (Rapp 1992) where I mention contributions made by my Swedish colleagues Rudberg (1977), Malmström and Palmér (1984), Lagerbäck (1988), Kleman and Borgström (1990), Holmlund (1991), and I now add Kleman (1992). The findings and interpretations made by King (1986) on altitudinal zonation of contemporary mountain permafrost in Scandinavia show that the lower limit of widespread, discontinuous permafrost is at a mean annual air temperature of about −4°C. That also means widespread permafrost above *ca.* 1200 m a.s.l. in the Abisko and Kebnekaise mountains in northern Lapland, somewhat lower on north-facing slopes. Wet-based ice eroded and removed the loose debris in glacial flow zones (cf. Kleman and Borgström 1990).

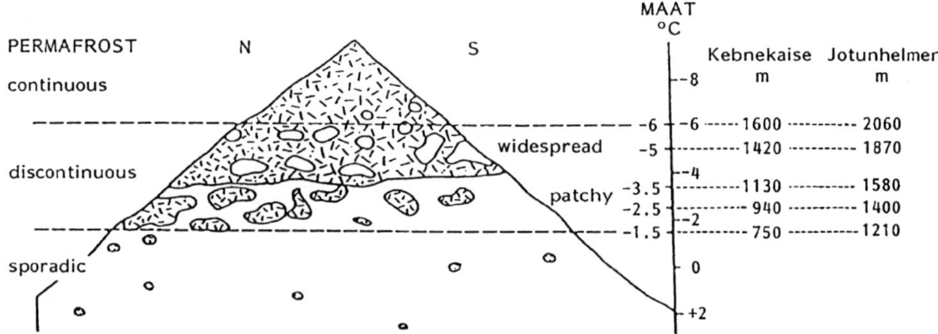

FIGURE 5.2 Altitudinal zonation of high mountain permafrost (L. King 1986). MAAT = mean annual air temperature. Transects from north to south in Kebnekaise, north Sweden, and Jotunheimen, south Norway

These arguments and local field observations and measurements of processes were the basis of my hypothetical sketches (Rapp 1992) of glacial and periglacial erosion in three altitudinal zones in the Abisko mountains (Figure 5.3). The uppermost zone above 1200 m is a zone of weak or zero glacial erosion (= non-scoured zone, Dyke 1993) mainly due to cold-based ice and weak periglacial action in cold and dry locations with mountain permafrost. An old morphology with tor-like forms in granite or similar rocks, or cryoplanation steps in layered bedrock is likely to survive glaciations. Permafrost is likely in this zone, above 1200 m a.s.l. Zone B is in the 600–1200 m level. It has moderate or locally strong glacial erosion and strongest periglacial action. Glacier cirques, overdeepened tarns and maybe also some nivation hollows are considered to be well developed in this zone. Wet-based ice conditions prevail. Zone C experiences strong

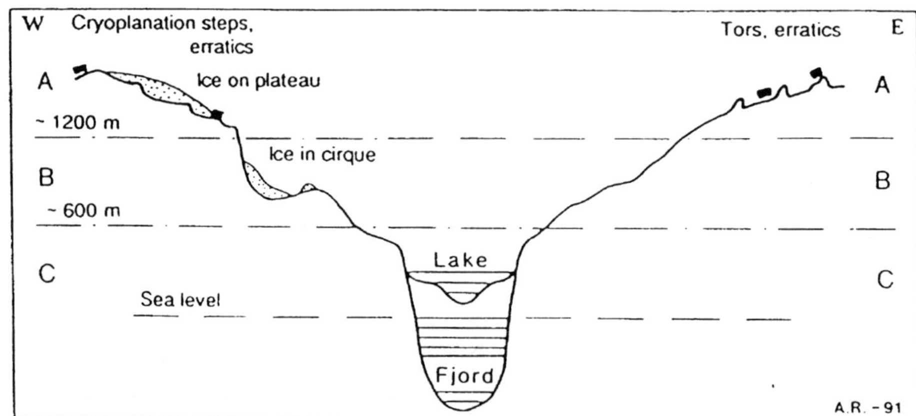

FIGURE 5.3 Hypothetical sketch of altitude zonation of glacial and periglacial erosion, based on field notes from Kärkevagge area (W) and the eastern mountains (E). See text for details. Drawing by A. Rapp

glacial erosion due to wet-based ice and confined ice flow during glacial maxima. This results in overdeepened basins in bedrock (fjords and lakes).

SUMMIT TORS CARRYING GLACIAL ERRATIC BLOCKS

Tors are pillars or knobs of bedrock some metres high which have formed by bedrock weathering and later removal of weathered material by creep and wash processes, not by waves or glacial action. In glacially scoured terrain the exposed bare bedrock is typically glacially eroded into whale-back rocks, also called roches moutonnées. They have stoss and lee sides and surfaces with glacial striations sometimes crossing systems of striae. Other signs of glacial scour which reflect the flow direction of scouring ice include crescentic fractures and crescentic gouges ("Sichelwannen" in German).

Well-developed tors occur in Europe south of the former glacial borders, e.g. in Dartmoor in south England, in Germany, south Poland and Czech Republic. But tors also occur in non-glaciated mountain tundra. I have photographed tors in quartzitic rocks near

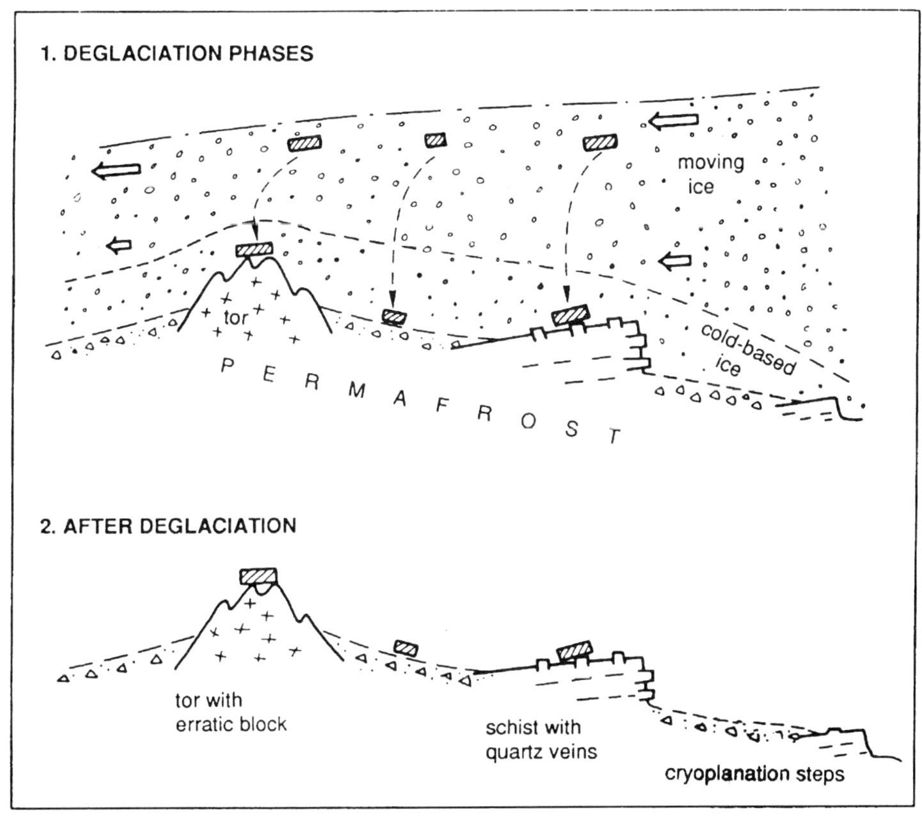

FIGURE 5.4 Hypothetical sketches of mechanism of inland ice moving over cold-based ice, protecting its bed of frozen blockfield from glacial scour. After deglaciation, bedrock steps and blockfields can remain untouched by glacial scour, but carry long-transported glacial erratic blocks. The ice-man of Tyrol found in 1991 (Patzelt 1992) supports this hypothesis. Drawing by A. Rapp

Klondike, Yukon, in combination with cryoplanation terraces. In many cases tors and cryoplanation steps are carrying long-transported erratic blocks. It means that non-scoured summit areas with tors, blockfields and cryoplanation steps have no traces of glacial striations, but have drastic forms of periglacial micro-weathering such as knobs and veins of quartz protruding as much as 50–100 mm above weathered micaschist siderock. They also have wide, open joints with rounded edges. Mt Vassitjåkka, west of Kärkevagge, Abisko, has all these indicators of non-scoured zones from *ca.* 1200 m a.s.l. to the summit plateau of 1590 m and also well-kept garnet crystals rising up to 50 mm above the rock surfaces without sign of glacial scour. But erratic blocks on top of the tors and on bedrock steps with protruding quartz and garnets indicate that the inland ice was there, probably moving over a ground-protecting crust of cold-based ice. I suggested the hypothesis of cold-based ice on permafrost in text and provide explanatory figures (Rapp 1992), shown here as Figure 5.4. I think this hypothesis of conserving cold-based ice beneath moving glacier ice is important to consider in connection with continued discussions on the nunatak problem. It is also supported by field observations from many excursions with colleagues, e.g. to Mt Aurivara, north Sweden, in 1984 and 1992 (gabbro and syenite tors); Disko Island, West Greenland, in 1985 (basalt cryoplanation treads and risers with erratics of gneiss); Gaspé peninsula, Quebec, in 1990 (Shick-shack mountains, serpentine tors with gneissic erratics). In Lapland's western mountains of garnet micaschist there are tor-like forms and cryoplanation steps with long-transported erratics of eastern, pink granites and dark amphibolites. The tors of, for example, Mount Kärketjårro east of Kärkevagge are angular cube-like ruins, like the tors in West Virginia's Appalachians in gently dipping orthoquartzite (cf. fig. 2.5 in Clark and Hedges 1992, p. 44). These forms are very different from the fresh or weathered ice-scoured whale-back rocks (roches moutonnées) which are typical of most valley bottoms and high cols where glacial scour was concentrated by rapid ice flow during the glaciations.

Field studies of 15 summit tors on Mt Aurivara, north Sweden, were performed by M-F. André in July 1993 (personal communication). A few of these had glacial striations on flat, exposed surfaces and most of the tors carried long-distance transported glacial erratic boulders. Some of them were dropped from above into open joints. We think this shows a weak glacial scour and proves that the joints are preglacial weathering joints. The tors of Mt Aurivara are the first to be interpreted and described as tors in the north Swedish mountains. (cf. Figures 5.5 and 5.6).

BLOCKFIELDS

Blockfields, also called felsenmeer, are areas covered by mostly angular blocks, carrying a poor vegetation of lichens, mosses and scattered tufts of grass or herb species. The plateaus of northern mountains are generally dominated by blockfields. "Normally, unlike scree, all these deposits lack a cliff at their head, and all have long attracted attention and stimulated much debate as to origin and environmental significance." (Washburn 1979). The mountain blockfields of north Scandinavia have been studied and analysed in monographs by, for example, R. Dahl (1963), L. Strömquist (1973) and Malmström and Palmér (1984). They all agree on a general thickness of about 1–1.5 m and a stratigraphy of blocks on top and smaller angular debris below, including silty

FIGURE 5.5 Summit tors and cryoplanation terraces on Mt Aurivare, Kiruna. These landforms have probably escaped heavy glacial erosion due to cold-based, non-erosive ice during their main glaciation. Bedrock is gabbro, altitude *ca.* 1000 m a.s.l. and location west of Rensjön village, 30 km northwest of Kiruna. Note rounded core-stones of tor in foreground and reindeer corral as bright area in the centre. Photograph A. Rapp

FIGURE 5.6 Summit tor of gabbro with rounded core-stones. Mt Aurivare, Kiruna. Photograph A. Rapp

material, and an abrupt contact with bedrock. O. Svenningsen (oral communication 1993) reported *in situ* weathered blockfields at 1800–2000 m a.s.l. in the Sarek mountains, with glacial erratics on the surface.

In a paper from the 1980s (Rapp 1986) I wrote the following comments of the blockfield and refugia discussion: The nunatak theory or hypothesis is considered by its defenders to be supported by the existence of strongly weathered "old" looking blockfields above a certain level ("trimline") in these mountains (Ives 1974). It is important that this discussion about the genesis of alpine blockfields continues for a better understanding of the Quaternary history of mountain vegetation, the extent of the glaciations, and the effects of glacial erosion on high levels. The rate of blockfield weathering is a key issue. R. Dahl (1967) concludes that the blockfields at high levels in north Norway have resulted from postglacial, local frost wedging with an addition of a few glacial erratics left by the Weichselian glaciation. The same interpretation is applied by Malmström and Palmér (1984) in their studies of blockfields in the Varanger Peninsula in northern Norway. Another possible interpretation might be that the ice movement at high levels and very cold conditions during the Weichselian maximum could have been so slow and erosion so weak from a "cold-based ice sheet", that a preglacial weathering mantle could have escaped removal by the ice and could form the main part of the debris in the present-day blockfields. The latter interpretation is similar to the one proposed by Sugden and Watts (1977) on blockfields in Baffin Island, Arctic Canada.

Eilif Dahl (1987) discussed the origin of so-called mountain-top detritus in Norway and Canada. Presence of gibbsite and other clay minerals suggest that the mountain-top detritus is a remnant of the Tertiary weathering crust which has survived the Pleistocene glacial ages. He interprets the mentioned clay minerals as an indicator of ice-free refugia, but I prefer to regard them as indicators of non-scoured zones covered by and protected by cold-based ice.

In the discussions of mountain blockfields and their genesis in northern Scandinavia three theories have been proposed:

1. Blockfields are mainly formed by postglacial weathering of 9000–13 000 years duration since the latest main deglaciation.
2. Blockfields are mainly formed by subglacial till.
3. Blockfields are mainly formed by preglacial weathering and can "survive" glaciations due to weak glacial scour, e.g. beneath cold-based ice.

Case 3 conditions have been described and used in interpretations in different environments in Scandinavia, e.g. till-covered woodlands in northeastern Sweden (Lagerbäck 1988), mountain blockfields on plateaus in central Sweden (Kleman and Borgström 1990) and in the Abisko mountains in north Sweden (Kleman 1992; Rapp 1992).

Case no. 1 and 2 above are ruled out and case no. 3 is supported by observations at mountain localities, where zones of bare, unweathered bedrock occur side by side with zones of local weathered blockfield of the same bedrock. Such key localities are wash zones from the main deglaciation *ca.* 9000 BP where lateral glacial drainage floods removed the blockfield cover in wide channels where no bedrock weathering happened later in the Holocene (Figures 5.7 and 5.8). Other key localities are high

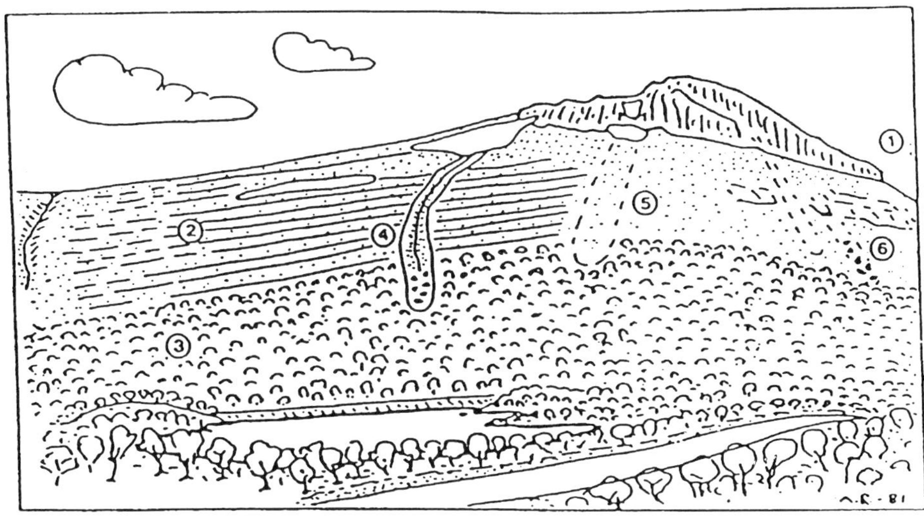

FIGURE 5.7 Drawing of Mt Kaisepakte seen from the north. 1. Cliff of overthrust sheet. 2. A system of lateral drainage channels from deglaciation floods. 3. Birch forest. 4. and 5. Tracks of erosive snow avalanches. 6. Recent rockslide. Drawing by A. Rapp 1981

passes or cols in the blockfield zone, e.g. in micaschist mountains where concentrated flow zones of glacial scour occur side by side with non-scoured ridges and plateaus with blockfield cover and sometimes also with tors and cryoplanation terraces and steps (Dyke 1993).

Micro-weathering forms of glacially scoured and non-scoured zones of cold mountains are useful for the diagnosis. Inventories of glacial striations, their occurrence and directions are fundamental for the understanding of glacial scour impact and history. Quantitative studies of the micro-weathering of bedrock surfaces have been done by several authors who measured the lowering of bedrock close to resistant veins or knobs of quartz. A diagram by R. Dahl (1967) on granite slopes of north Norway shows zero weathering at sea-level, maximum of 10–15 mm at the highest coastline at *ca.* 100 m a.s.l. and low values below 4 mm at the elevations of the blockfield zone *ca.* 1200–1500 m altitude. Such studies are being performed now by Dr M-F. André.

Typical weathering forms of the non-scoured alpine blockfield zone of north Sweden at 1200–1700 m altitude are quartz knobs of up to 50–100 mm relief, selectively weathered crystals of garnet in garnet micaschists, wide weathering joints with rounded edges, patterned ground and stone stripes appearing on newly exposed land after melting of earlier ice cover (cf. overview map, Figure 5.10).

COLD-BASED GLACIERS AS ENVIRONMENTAL ARCHIVES ON BLOCKFIELDS

Of particular importance are observations at supposed cold-based mountain glaciers, which are melting today, due to greenhouse warming or other reasons. What interpretations can we make from such studies about their scouring or non-scouring, conservational effects?

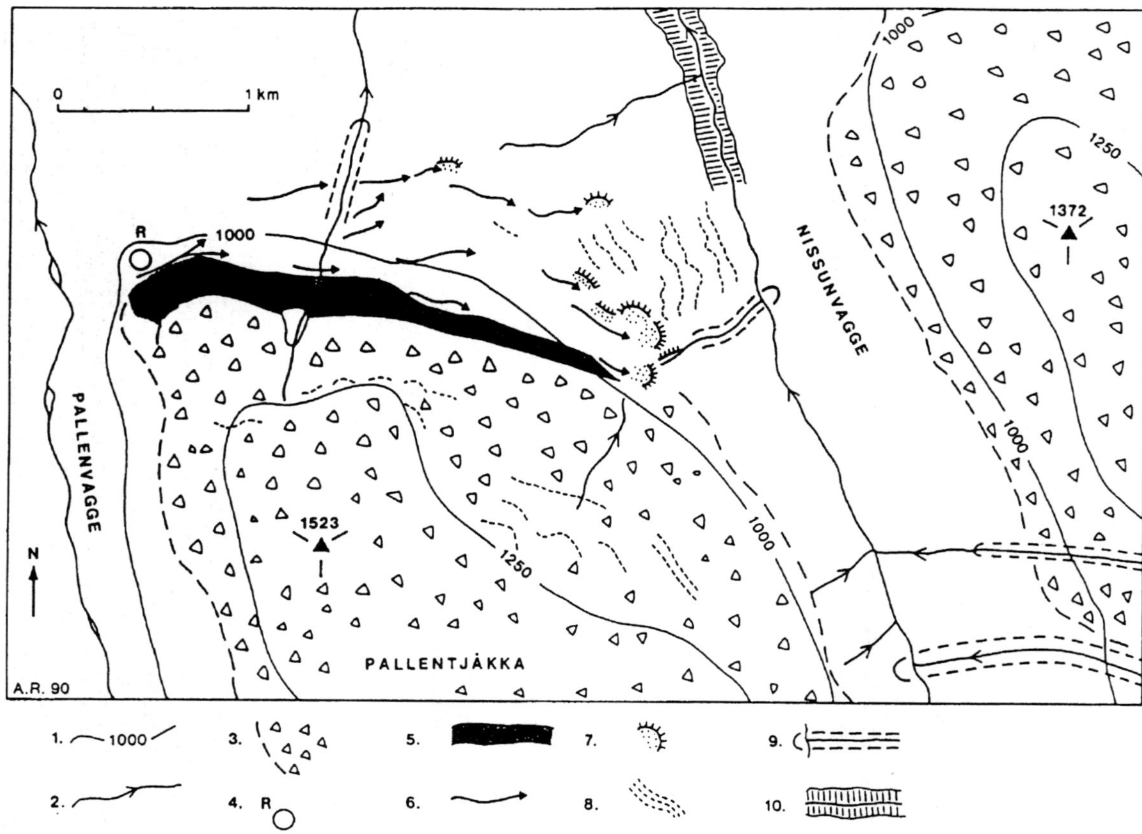

FIGURE 5.8 Alpine blockfields and glaciofluvial wash zone, Abisko. Drawing by A. Rapp. 1. Contour line. 2. Stream and flow direction. 3. Blockfield and its lower limit. 4. Reindeer corral. 5. Wash zone of deglaciation flood. 6. Lateral drainage channel. 7. Glaciofluvial delta terrace. 8. Deformed lateral drainage lines (?). 9. Slush avalanche track. 10. Canyon at Nissunjåkka

An area which offers very favourable opportunities to study present-day plateau glaciers and their mechanisms, scouring and geomorphological impact on blockfields, lakes and lake sediments, is the western Abisko mountains, with the Låkta plateau and the mountains on both sides of the Kärkevagge valley. It is a reference area of international importance for studies of alpine and periglacial slope processes and resulting landforms. Rapp (1960) studied slope processes in the lower and middle altitude zones of Kärkevagge. Current projects on nivation and other studies elsewhere have highlighted problems concerning processes in the higher zone. A new project aims at extending earlier process studies in Kärkevagge to the high plateaus with blockfields, snowfields and a melting plateau glacier probably with cold-based ice on permafrost. The Ekman glacier at 1235–1400 m a.s.l. south of Låkta will also be used for similar studies (Figures 5.9 and 5.11).

Lake Rissajaure in Kärkevagge is considered to be the clearest lake in Europe with 34 m transparency measured several times. It receives the meltwater from the plateau glacier on Mt Vassitjåkka, but very little glacial silt is nowadays carried by the meltwater from the glacier to Lake Rissajaure. This is in drastic contrast to the melting cirque glacier nearby, which produces very turbid water during periods of rapid melting in summer. It has a large colluvial fan and large delta deposits in the valley.

A thesis by Jonasson (1991) presents data on sediment stratigraphy and dating of a 2.7 m-long sediment core from Lake Rissajaure (Figure 5.11). The sedimentation rate, as described by interpolations between three ^{14}C datings has been quite even from *ca.* 7300

FIGURE 5.9 Front of the Ekman glacier and lake. Photograph B. Holmgren, 15 September 1985

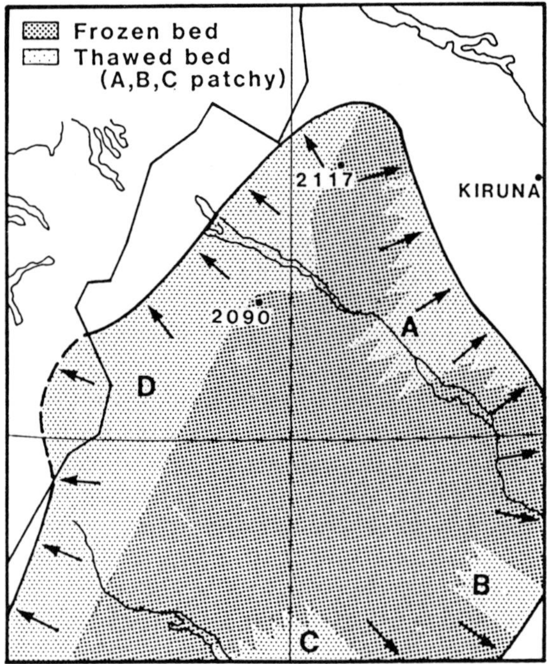

FIGURE 5.10 Basal thermal zonation and approximate flow pattern during one stage of the final Late Weichselian recession. The approximate extent of frozen-bed conditions is reconstructed on the basis of preserved deposit landforms from older stadials. Spot heights indicate the highest points in the Kebnekaise and Sarek massifs (Kleman 1992)

BP to *ca.* 2000 BP. The events of heavy mass movements and sedimentation have occurred mainly during the last 4000 years (Jonasson 1991). It is likely that there were no glaciers in the catchment during the Holocene climatic optimum *ca.* 9000–5000 BP, but that they grew and disappeared in several episodes later (e.g. in the Little Ice Age, from *ca.* 1550 to heavy melting after 1920).

Examples of cold-based glaciers preserving vegetation can be quoted from other areas, e.g. the Canadian Arctic.

Blake (1981) reported numerous finds of intact *Salix* spp., *Dryas integrifolia* and moss specimens from 1000 to 5500 years old on Ellesmere Island. In a paper by Bergsma et al. (1984) the authors reported on frozen and dead but well-preserved arctic plant communities conserved under glacial ice at Irwin Glacier, Ellesmere Island. The plants

---

FIGURE 5.11 (*opposite*) Map of the Kärkevagge area. Key: 1. Contour lines, 20 m equidistance. 2. Limit of block zone deposits in valley bootom. 3. Giant blocks. 4. Morainic ridges. 5. Lateral drainage terraces. 6. Solifluction slopes. 7. Major wind direction of snow drifting. Full-drawn line = cirque headwall. Dashed line = nivation hollow? 8. Glacial striations. 9. Glacier area in 1943. A–D, H–L, V = valley-side sections

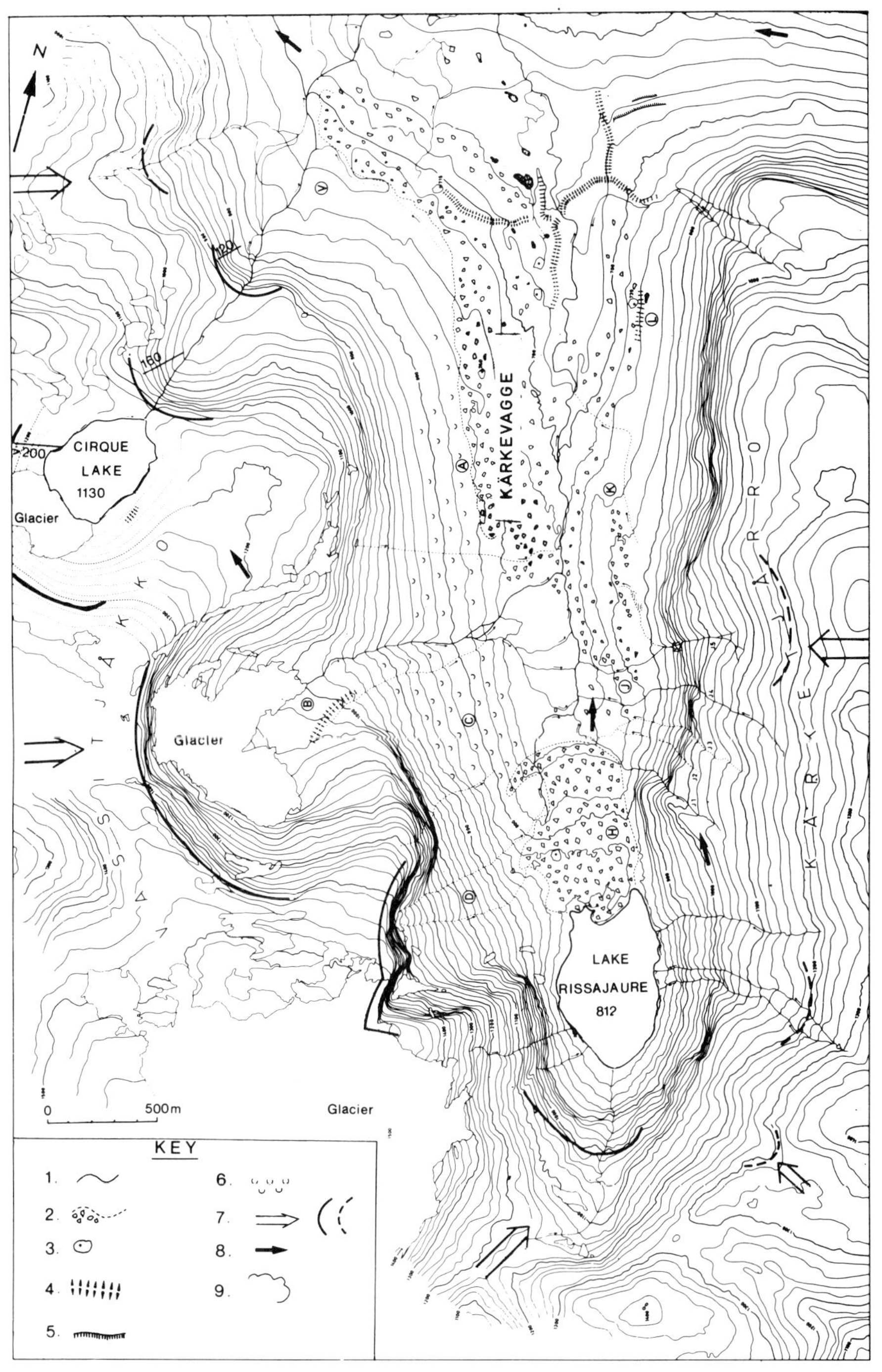

Figure 5.11 *For caption see opposite*

of a *Cassiope–Dryas* community were dated as about 400 radiocarbon years old: "The excellent preservation of the plants supports the thesis that polar glaciers are frozen to their bases, and hence their movements are by internal deformation rather than by erosive basal sliding." (Bergsma et al. 1984).

The most drastic example of how cold-based ice can preserve organic material is the find of the so-called Iceman Ötzi in September 1991 near the Similaun pass in the Alps of Tirol in Europe. The mummified body of an intact man from the stone age, dated to 5300 years BP, was found with his equipment and clothes resting on a blockfield between two ridges of bedrock *ca.* 2–3 m high (Patzelt 1992). The find of Ötzi indicates that a local Neoglaciation began in the Alps at the end of the Atlantic warm period and lasted to our time. The largest glacier in Scandinavia, Jostedalsbre in west Norway, began to exist at the same time (Nesje and Kvamme 1991; Karlén and Matthews 1992). In mountain ranges with cold-based glaciers, scientists should be prepared to investigate new finds of glacially non-scoured landforms, soils and frozen vegetation, which may be exposed if climatic warming continues.

ACKNOWLEDGEMENTS

I am greatly indebted to many colleagues for stimulating encouragement, and help with field research and excursions in the Abisko area from 1952 onwards. The Natural Science Research Council, the Royal Academy of Sciences and the Crafoord Foundation gave financial support to the international excursions of 1988, 1990, 1991 and 1992. Professors T.V. Callaghan and Sven Jonasson at Abisko helped with references on plant ecology. Mrs B. Fogelström and Mrs E. Särbring, Lund, did the typing and drawing. My warm thanks to these persons and institutions.

REFERENCES

Åkerman, J. and Malmström, B. 1986. Permafrost mounds in the Abisko area, Northern Sweden, *Geografiska Annaler*, **68A**, 155–166.

Bergsma, B.M., Svoboda, J. and Freedman, B. 1984. Entombed plant communities released by a retreating glacier at Central Ellesmere Island, Canada. *Arctic*, **37**(1), 49–52.

Blake, W. Jr. 1981. Neoglacial fluctuations of glaciers, southeastern Ellesmere Island, Canadian Arctic. *Geografiska Annaler, 63A*, 201–218.

Clark, G. M. and Hedges, J. 1992. Origin of certain high-elevation local broad uplands in the central Appalachians, USA. In Dixon, J.C. and Abrahams, A.D. (eds), *Periglacial Geomorphology*. Binghamton Symposia, 22. Wiley, Chichester, pp. 31–62.

Dahl, E. 1987. The nunatak theory reconsidered. In Research in Arctic life and earth sciences: present knowledge and future perspectives (Editor: M. Sonesson), *Ecological Bulletins*, **38**, 77–94, Copenhagen.

Dahl, R. 1963. Shifting ice culmination, alternating ice covering and ambulant refuge organisms. *Geografiska Annaler*, **45A**, 122–138.

Dahl, R. 1967. Post-glacial micro-weathering of bedrock in the Narvik district, Norway. *Geografiska Annaler*, **49A**, 155–166.

Dyke, A.S. 1993. Landscapes of cold-centred Late Wisconsinan ice caps, Arctic Canada. *Progress in Physical Geography*, **17**(2), 223–247.

Faegri, K. 1993. Overvintringssymposium i Bergen. *Blyttia*, **2**, 70–72.

Holmlund, P. 1991. Cirques at low altitudes need not necessarily have been cut by small glaciers. *Geografiska Annaler*, **73A**, 9–16.

Hoppe, G. 1959. Några Kritiska Kommentarer till distkussionen om isfria refugia. *Svensk Naturvetenskap, Årsbok*, 123–134.

Hoppe, G. 1983. *Fjällens terrängformer. Statens Naturvårdsverk. Rapporter*. Liber, Stockholm, 60 pp.
Hultén, E. 1950. *Atlas of the Distribution of Vascular Plants in NW Europe*. Esselte, Stockholm, 512 pp.
Ives, J.D. 1974. Biological refugia and the nunatak hypothesis. In Ives, J. and Barry, R. (eds), *Arctic and Alpine Environments*. Methuen, London, 999 pp.
Jonasson, C. 1991. Holocene slope processes of periglacial mountain areas in Scandinavia and Poland. *Uppsala University, Department Physical Geography, Rapport*. **79**, 156 pp.
Jonsson, S. 1983. On the geomorphology and past glaciation of Storöya, Svalbard. *Geografiska Annaler*, **65A**, 1–17.
Kaitanen, V. 1989. Relationships between icesheet dynamics and bedrock relief on dissected plateau areas in Finnish Lappland. *Geografiska Annaler*, **71A**, 1–15.
Karlén, W. and Matthews, J.A. 1992. Reconstructing Holocene glacier variations from glacial lake sediments: Studies from Nordvestlandet and Jostedalsbreen–Jotunheimen, southern Norway. *Geografiska Annaler*, **74A**(4), 327–348.
King, L. 1986. Zonation and ecology of high mountain permafrost in Scandinavia. *Geografiska Annaler*, **68A**, 131–139.
Kleman, J. 1992. The palimpsest glacial landscape in northwest Sweden. *Geografiska Annaler*, **74A**, 305–326.
Kleman, J. and Borgström, J. 1990. The boulder fields of Mt. Fulufjället, west-central Sweden. *Geografiska Annaler*, **72A**, 63–78.
Lagerbäck, R. 1988. Periglacial phenomena in the wooded areas of northern Sweden — relicts from the Tärendo interstadial. *Boreas*, **17**, 487–500.
Malmström, B. and Palmér, O. 1984. Glacial and periglacial geomorphology on the Varanger Peninsula, northern Norway. *Meddelanden Lunds Universitets Geografiska Inst., Avhandlingar*, **93**, 351 pp.
Nesje, A. and Kvamme, M. 1991. Holocene glacier and climate variations in western Norway. *Geology*, **19**, 610–612.
Nyberg, R. 1985. Debris flows and slush avalanches in northern Swedish Lappland. *Medd. Lunds Geogr. Inst., Avh.*, **97**, 222 pp.
Nyberg, R. 1989. Observations of slushflows and their geomorphological effects in the Swedish mountain area. *Geografiska Annaler*, **91A**(3–4), 185–198.
Nyberg, R. 1991. Geomorphic processes at snowpatch sites in the Abisko mountains, northern Sweden. *Zeitschrift für Geomorphologie*, **35**, 321–343.
Nyberg, R. and Lindh, L. 1990. Geomorphic features as indicators of climatic fluctuations in a periglacial environment, northern Sweden. *Geografiska Annaler*, **72A**(2), 203–210.
Österholm, H. 1988. The glacial striation on Prins Oscar's Land, Nordaustlandet, Svalbard. *Univ. of Stockholm, Naturgeografisk Forskningsrapport*, **70**, 10 pp.
Patzelt, G. 1992. Neues vom Ötztaler Eismann. *Österreichischer Alpenverein-Mitteilungen*, 2/92.
Porsild, A.E., Harington, C.R. and Mulligan, G.A. 1967. Lupinus arcticus (Wats.) grown from seeds of Pleistocene age. *Science*, 158, 113–114.
Rapp, A. 1960. Recent development of mountain slopes in Kärkevagge and surroundings, Northern Scandinavia. *Geografiska Annaler*, **42**, 71–200.
Rapp, A. 1986. Slope processes in high mountains. *Progress in Physical Geography*, **10**(1), 53–68.
Rapp, A. 1992. Kärkevagge revisited. Field excursions on geomorphology and environmental history in the Abisko mountains, Sweden. *Sveriges Geol. Undersökning, Ca.*, **81**, 269–276.
Rapp, A., Nyberg, R. and Lindh, L. 1986. Nivation and local glaciation in North and South Sweden. *Geografiska Annaler*, **68A**, 197–205.
Rudberg, S. 1977. Periglacial zonation in Scandinavia. In Poser, H. (ed.), *Abhandlung der Akademie der Wissenschaften in Göttingen*, Folge 3, no. 31, pp. 92–104.
Seppälä, M. 1986. *Detailed geomorphological maps. Atlas of Finland*. Folio 121–122. Nat. Board of Survey. Helsinki.
Sollid, J.L., 1988. Excursion guide. North Scandinavia. *5th International Conference on Permafrost*. Trondheim. (Mimeo.)
Strömquist, L. 1973. Geomorfologiska studier av blockhav och blockfält i norra Skandinavien.

UNGI Rapporter, **82**, 161 pp, Uppsala.

Sugden, D. and Watts, S.H. 1977. Tors, felsenmeer and glaciation in northern Cumberland Peninsula, Baffin Island. *Canadian Journal of Earth Sciences*, **14**, 2817–2823.

Washburn, A.L. 1979. *Geocryology*. E. Arnold, 406 pp.

Watts, S.H. 1983. Weathering processes and products under arid arctic conditions. *Geografiska Annaler*, **65A**, 85–98.

6

Karst in a Cold Climate: Effects of Glaciation and Permafrost Conditions Upon the Karst Landform Systems of Canada

DEREK C. FORD

Department of Geography, McMaster University, Hamilton, Canada

INTRODUCTION

It was a privilege and a pleasure to host the Third International Geomorphology Conference at McMaster University in Hamilton, Ontario. When the chairman of the Conference programme committee, Professor Olav Slaymaker, invited me to represent our local hosting group of geomorphologists in the series of plenary lectures that he proposed, it seemed appropriate to select a topic that would show something of the very rich diversity of Canada to geomorphologists from other nations. My specialization is in karst geomorphology and caves: "Karst in a cold climate" utilized some 120 colour-slide scenes to illustrate landscapes and karst features in nine of the provinces of Canada and in its Northwest Territories. To link the many different places in this geographical excursion together, the theme of the impact of glaciation upon karst landforms and processes was adopted. It is a truism that much of Canadian physiography remains dominated by the effects of past Quaternary glaciers and ice sheets. The other geomorphologies can be viewed as reworking glacial erosional sculpture and glacial deposits, only to be pushed aside as the glaciers return. Karst geomorphology is something of an exception in this scenario, however; its guts, systems of cave conduits, can survive beneath the ice and even enjoy accelerated development there. The potential complexity of the situation, and the added complication of permafrost growing and shrinking in the north in counterpoint to the glacier cycles, is the subject of this essay. Most of the material in it has appeared previously in specialized texts and journals. It is presented here in the hope that readers will find it interesting to compare its content and perspective with those of the other contributors to this volume.

Geomorphology Sans Frontières. Edited by S. Brian McCann and Derek C. Ford.

THE KARST GEOMORPHIC SYSTEM

The diagnostic characteristic of the karst geomorphic system is the dominance of one process — the aqueous dissolution of a rock. In a karst, dissolution either is quantitatively predominant or it serves as the essential trigger process that permits other processes to operate, e.g. when the roof of a dissolutional cavern collapses mechanically.

Figure 6.1 is a cartoon intended to suggest much of the nature and content of the karst geomorphic system. Given the predominance of the process of dissolution, the outstanding feature is that water is circulated underground via systems of solution-enlarged fissures or conduits (caves) instead of discharging through surface river channels. Being sufficiently deep, most conduit systems may survive destruction by scouring glaciers whereas surface channels will not. Surface river networks must be laid out anew after each glaciation but underground systems can survive in a modified form. Although Figure 6.1 (being comprehensive) indicates that karst channel networks may be constructed by expelled thermal and other deep waters, most networks in Canada have been created by meteoric waters circulating to depths of no more than a few hundred metres along flow paths ranging from hundreds of metres to several tens of kilometres in their length.

The distribution of karst rocks in Canada is given in Figure 6.2. For gypsum and anhydrite, limestone and dolomite only the "outcrop" is shown, i.e. where these rocks are not overlain by other consolidated rocks, although they may be mantled by veneers of glacial detritus, etc. These outcrops amount to approximately $1{\cdot}2 \times 10^6\,km^2$ in the aggregate. In addition, there is an area at least $0{\cdot}7 \times 10^6\,km^2$ beneath Quebec, Ontario, the Prairie Provinces and the Mackenzie Valley where these rocks are overlain only by a shallow cover of later consolidated rocks and so may be subject to modern interstratal dissolution. Salt is present at depths of 200–2400 m beneath $0{\cdot}5 \times 10^6\,km^2$ of the Prairie Provinces and in much smaller amounts in Nova Scotia, New Brunswick, Ontario, British Columbia, the Mackenzie Valley and the arctic islands. It will dissolve wherever groundwater can be brought to it.

The karst rocks occur in every tectonic and topographic setting known in Canada. Flat-lying and little-deformed spreads of platformal limestones, dolomites and gypsum are predominant, covering large areas in southern Ontario, the Hudson Bay Lowlands, Manitoba, the Mackenzie Valley and the southern arctic islands. Highly deformed masses occur in the Appalachian Orogen in western Newfoundland, Nova Scotia and the Gaspé Peninsula of Quebec. Deformed and often metamorphosed carbonates occur in the Canadian Shield, especially in the Grenville geologic province of Ontario and Quebec (marbles) and the Labrador Trough. Great thrust sheets of limestone and dolomite are the principal mountain-building rocks in many ranges of the Rocky Mountains and the Mackenzie Mountains.

Given such an abundance of suitable rock plus climates that were of the humid to subhumid types, it must be supposed that karst landforms existed before the first glaciers spread across Canada. Karst processes will have operated during all succeeding interglacials and interstadials; to a more limited extent and in a somewhat different manner they may even have operated beneath glaciers, as described below. There has been much karst–versus–glacier process competition and also karst–glacier process interaction. This ensures that the modern Canadian karst will have a complex inheritance.

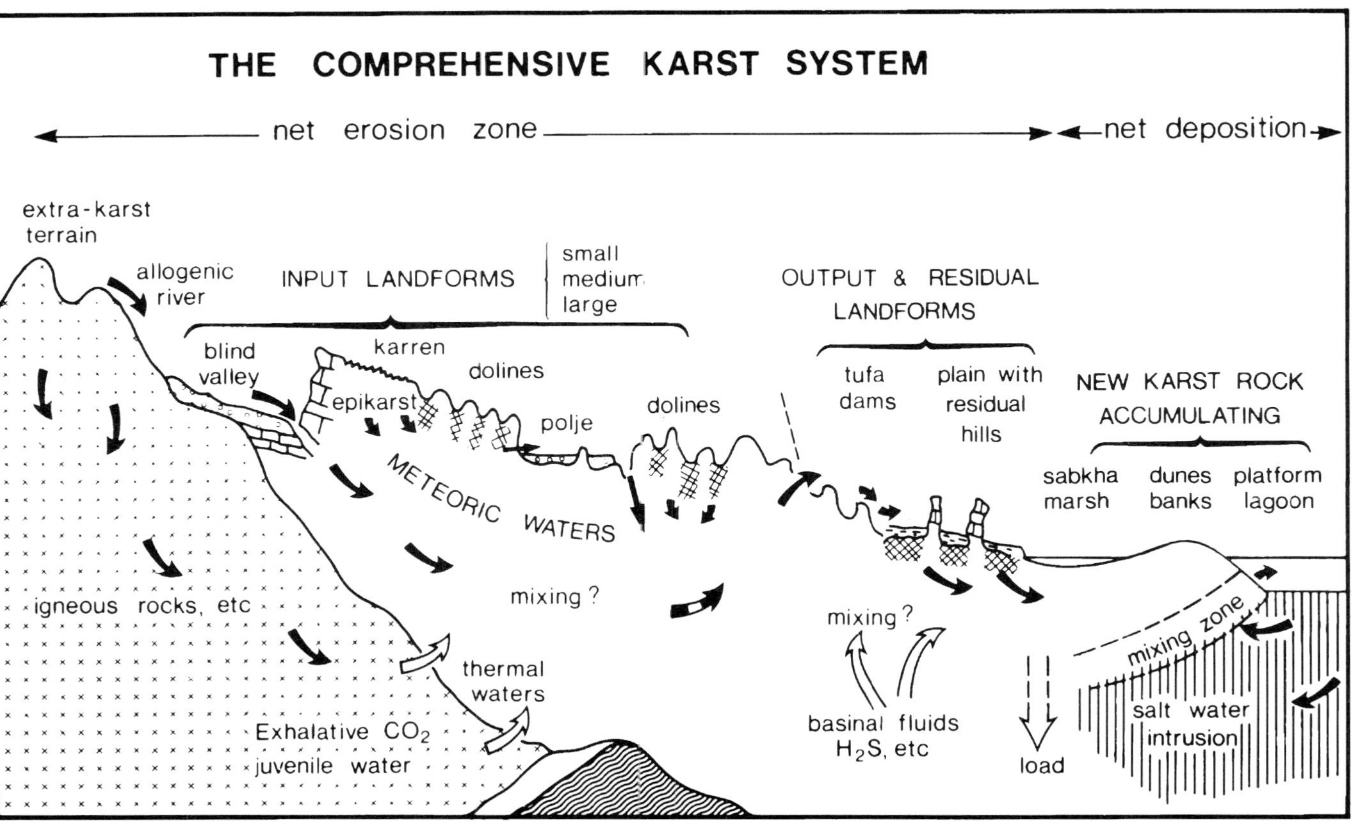

FIGURE 6.1 The nature and content of the karst geomorphic system

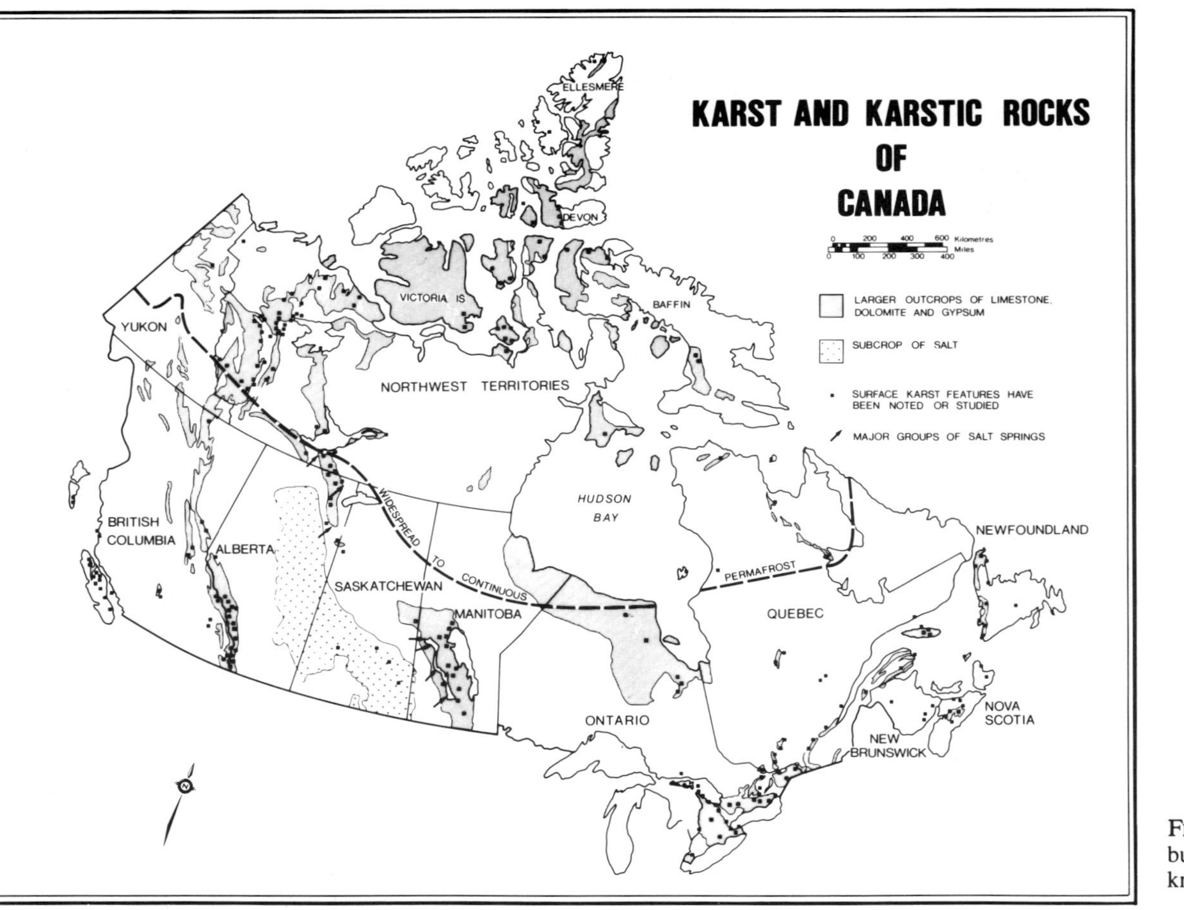

FIGURE 6.2 The distribution of karst rocks and known karst landforms in Canada

In Figure 6.2 it is seen that karst rocks are widespread in the modern permafrost regions of Canada. Permafrost must impose limitations upon the groundwater circulation that is a necessity for karst system development. However, most of the presently permafrozen karst outcrops were ice-covered during the major glaciations. The ice then insulated them, perhaps permitting the permafrost to thaw. Interactions between karst and glacial and permafrost processes, simultaneously or in sequence, may be especially complicated. They are addressed in a later section of this review.

EFFECTS OF GLACIERIZATION UPON KARST SYSTEMS

Table 6.1 groups and lists the nine distinct effects that I have found may occur when glaciers override previously established karst landforms and karst aquifers (groundwater flow networks). There may be further effects awaiting discovery, although the range of possibilities cited here appears to be quite comprehensive in its embrace.

Erasure of shallow karst features by glacier basal plucking and scour is a widespread and important effect. It implies that the glacier is wet-based so that it can slide over the bed, perhaps freezing down and plucking where the lubricating water film can drain away quickly into karst groundwater channels. Erasure impacts particularly upon microkarren and karren: because they are small they are fragile and tend to be removed entirely. Where high densities of them have created a limestone pavement (or "clint-and-grike" terrain — Sweeting 1972) this will be largely stripped away.

On many pavements erasure is not complete, however. In any one glaciation the ice appears to be capable of removing only the top one or two limestone beds (~ 1–2 m of strata). Bases of grikes (solution troughs) that extend down the deeper master joints in an area thus tend to be preserved. They may then guide the development of a new generation

TABLE 6.1 Effects of glacier action upon karst systems

Destructive, deranging;

1. Erasure — of karren and other small karst landforms
2. Dissection — of integrated systems of conduits
3. Infilling — of sinkholes and larger input features; aggradation of springs
4. Injection — of clastic detritus into cave systems

Inhibitive:

5. Shielding — carbonate- or sulphate-rich drift protects bedrock surfaces from postglacial solution

Preservative:

6. Sealing — clay-rich deposits seal and confine epikarst aquifers

Stimulative;

7. Focusing inputs, raising hydraulic head — with superimposed glacial streams or aquifers
8. Lowering spring elevations — by glacial entrenchment
9. Deep injection — of glacial meltwaters/groundwater when bedrocks are being flexed during crustal rebound or depression?

FIGURE 6.3 (a) To illustrate *glacial erasure* of karren, infilling and *postglacial shielding* of bedrock. This picture shows the base of a grike (kluft) in a dolostone quarry near Hamilton, Ontario. Wisconsinan ice scoured off the upper 1–2 m of rock, i.e. the grike walls. The surviving base was infilled and the scoured surface buried by 1·5 m of glacial till (which has just been cleared for bedrock quarrying). The scoured surface retains a pristine "glacial polish " 14 000 years after ice recession.

of karren after the ice has receded (as shown by grike width statistical studies on glaciated pavements in northern England prepared by Rose and Vincent 1986), or they may be rendered inert by the injection or shielding effects mentioned below. A few active, and many inert, beheaded grikes of this kind are exposed in the walls of limestone and dolostone quarries in Quebec, Ontario and Manitoba (Figure 6.3a).

There are few residual pinnacles or larger rock towers in the Canadian karst. Most that I know appear to be of postglacial age, the greatest being in areas of the remote (northwestern) Mackenzie Mountains that have not been glaciated for at least 350 000 years (Brook and Ford 1978). The implication is that glacial scour has removed earlier towers in such areas. Some larger limestone conical hills survive with ice-modified form in the Goose Arm area of Newfoundland and in the Benson River valley on Vancouver Island.

Glacial dissection of karst aquifers can occur because, in contrast to allogenic rivers, glaciers that invade a mature karst cannot be swallowed into its underground drains. Dissection has been most important in the alpine terrains, principally the Rocky Mountains. The finest example yet studied is on the Continental Divide at Crowsnest Pass, Alberta–British Columbia. Strata there form a west-dipping limestone cuesta. There is a central shale formation that is generally impermeable, dividing the limestone into

FIGURE 6.3 (b) To illustrate *glacial preservation* of karst. This scene was taken in a quarry at Bird's Hill, near Winnipeg, Manitoba. A clint-and-grike pavement is preserved intact benearth 2 m of melt-out till in this scene. In contrast to the situation of Figure 6.3a, the upper dolomite beds have not been scoured away despite being dissected by grikes and opened bedding planes

separate aquifers above and below it (Figure 6.4). Relict fragments of phreatic (sub-watertable) caves are scattered in the rock up to the highest summits, where they are certainly some millions of years old (Ford et al. 1981). From them it appears that early drainage in the cuesta was fully karstic and quite simple, being directed toward the Pass along the strike (Figure 6.4A). The cuesta has since been dissected by glacial cirques on the scarp face and by trough head valleys on the dip slope. As a consequence the modern drainage is a mixture of local surface elements and karst elements (Figure 6.4B). Strike-oriented underground drainage is preserved close to the Pass itself. On the north side this resurges in a major spring 6 m above Crowsnest Lake that is readily visible from the highway through the Pass. On the south side it includes a breach of the Continental Divide itself; water sinks in British Columbia and resurges at springs in Alberta. Further from the Pass some streams remain above ground, others sink to flow down-dip to the west or counter-dip to the east, while yet others are captured underground along the strike to pass from one cirque to another: the local situation evidently varied with the vagaries of glacier entrenchment interacting with solution caves that were established either subglacially or during earlier interglacial episodes.

Infilling describes the filling of karst input or output landforms by glacial or proglacial detritus. Glaciers can produce a great deal of rock debris. South of the city of Saskatoon closed depressions in bedrock as great as 12 × 25 km became wholly infilled by till and glacial lake deposits, with the result that they cease to exist as topographic entities

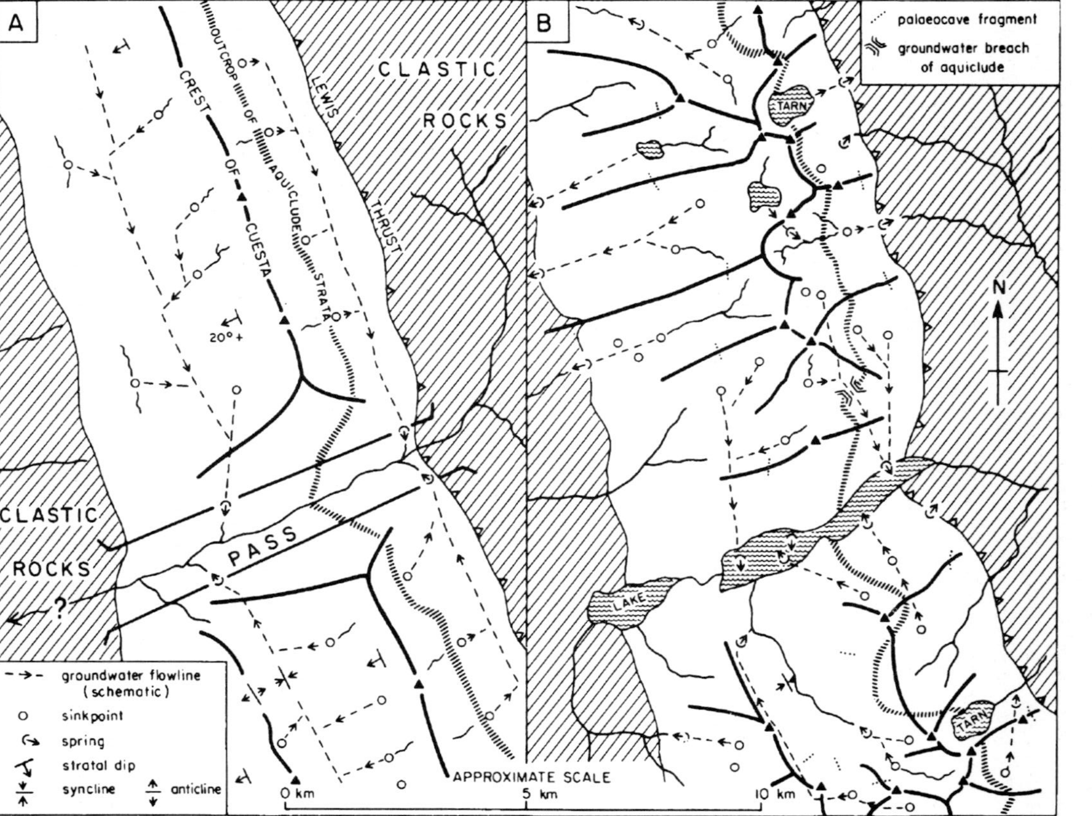

FIGURE 6.4 Schematic maps to display the dissection of paleokarst aquifers at Crowsnest Pass, Alberta, British Columbia. Before glaciation (Figure 6.4A) the aquifers were probably organized to give regional underground drainage to the Pass along the strike of the strata, one aquifer below (east of) and one above (west of) a shale aquiclude. Glacial dissection (Figure 6.4B) has fragmented the systems; underground, strike-oriented drainage is preserved only near the Pass

(Christiansen 1967). We may suppose that many thousands of lesser (doline-scale) karst depressions have been filled and lost in this manner in Canada.

Filling and burial do not necessarily imply that a doline or polje will be unable to perform its hydrologic role after the ice has receded. If the fill is permeable and there is a sufficient hydraulic gradient between sinkpoint and spring, water may pass through it into the aquifer. Given sufficient time the feature then may be re-excavated, partly or wholly. The great Canadian example is the Maligne River karst system, close to Jasper, Alberta. After draining 900 km^2 of mountain, forest and glacier the modern river sinks in a shallow basin in outwash sediments. The basin is seasonally flooded and known as Medicine Lake. Attempts to plug the lake-bottom sink artificially (to conserve the fishing) have always failed. The water resurges 16 km distant, 400 m lower in elevation, through more than 60 young springs in the floor of Maligne Canyon. The canyon is of postglacial age: at its mouth it is graded to the surface of a deep glacial infilling in a trunk valley. Flow-through time (sink to springs) may be as little as 12 hours (Brown 1972; Smart 1988). The modern Medicine Lake basin is 6 km in length and 0·8–1·5 km wide. I interpret it as an incomplete re-excavation of a greater karst polje that was infilled during the last glaciation and possibly earlier ones. Glaciation clogged both ends of the underground system but it quickly re-established itself; during the postglacial period at least $1{\cdot}8 \times 10^8\ m^3$ of glacial detritus has been cleared from the polje into the aquifer. Yet the latter continues to pass groundwater very rapidly indeed. Medicine Lake and Maligne Canyon are spectacular features that are readily accessible by car from Jasper.

Injection implies not merely the filling of the input or output landforms but the insertion of material deep into the conduit aquifer that lies between them. That is not so simple to accomplish but many examples have been found in Canada. The transporting streams may be supraglacial or englacial in position but in the majority of studied cases appear to have been subglacial or proglacial.

The most emphatic examples of injection are found in Castleguard Cave, Banff National Park, Alberta. This cave, the longest and finest yet known in Canada, now has its headwater passages beneath 200 m or more of glacier ice in the Columbia Icefield (see Ford 1983a). Seven of those passages are plugged by seals of glacier ice itself (Figure 6.5). This has been squeezed in like toothpaste. Probably the plugs are not very long (a few metres of injection only) but they are airtight and watertight. Castleguard Cave makes a deeper penetration beneath an extant glacier than any other cave that is known. Its ice plugs are unique.

Nakimu Caves, British Columbia, provide excellent examples of more commonplace injection. Their warren of conduits was filled to a depth of 120 m during the advance of the last glacier. The filling is a model for an ideal injection facies. At the base is a chaotic or sliding-bed deposit of the kind that has been described from eskers (e.g. Saunderson 1977). This is succeeded by sequences of well-sorted gravels and sands that fine upwards to terminate in 12 m of varved clays. The latter are interpreted as representative of full glacial conditions, when the karst aquifer was subordinate to water flow in a deep temperate glacier aquifer (Ford 1979). Much of the clastic filling was swept out again during the late glacial and early postglacial periods. Local water pressures and velocities were then at times so great (as evidenced by 1·5 m-diameter quartzite erratics wedged deep in the cave) that the hydrodynamics can be likened only to the flushing of a giant, glacier-primed, toilet bowl. It appears, therefore, that injecta into mountain aquifers will

FIGURE 6.5 Two differing perspectives on glacier injecta into caves. (a) Glacier ice from the base of the Columbia Icefield, Rocky Mountains, is viewed from a gallery into which it is intruding at the head of Castleguard Cave. (b) Glacial erratics injected into the entrance of a small cave in the Castleguard karst. Neoglacial ice receded from the photographer's position about 30 years before the picture was taken

rarely provide permanent and complete obstruction because the available flushing energy is too high.

One of the most peculiar karst lands of Canada lies between Goose Arm, Penguin Arm and Bonne Bay Big Pond in western Newfoundland. It is developed in folded, faulted limestones and dolomites and is a rugged terrain with a relief of 50 to 350 m. There are examples of mogote-form hills that appear to have been scoured into an asymmetric form by ice action. There are more than 40 large closed depressions that have also been modified by ice scour and/or till deposition (Figure 6.6). Most depressions contain lakes: some of these drain exclusively via underground channels, some exclusively by surface channels, while others drain underground but with seasonal (spring thaw) overflow. The three types are distributed without a discernible geographical pattern (Figure 6.7). The karst water chemistry and spring water temperatures are also unusual (Karolyi 1978). The truly peculiar feature, however, is that the modern underground drainage routes are small in scale, primitive or youthful in character and highly disorganized. There are nearly as many springs as there are dolines. Yet the large dimensions of the latter would demand, in a conventional extraglacial karst, that there be a well-integrated aquifer comprised of large conduits draining to just a few regional springs.

The surrounding terrains are granitic and metasedimentary highlands that supplied much clay- rich till to the karst. It blankets the depressions and interfluves. We believe that subglacial injection of the clay has fully clogged a previous integrated aquifer, causing partial inundation of the dolines and the development of the modern mixture of surficial and underground drainage routes that are postglacial in age (Karolyi and Ford 1983).

The renewal of dissolutional attack upon a karst bedrock following glacier recession may be inhibited if it is overlain by a detrital cover that contains fragments of the rock. The fragments absorb the attack neutralizing the water before it has descended as far as the bedrock. The effect is termed *shielding* (Table 6.1). It is very important, particularly in lowland Canada. In Ontario and Quebec we find that 1–2 m of permeable till or outwash with local clasts often has sufficed to protect underlying limestone or dolomite completely during the 11 000–14 000 years since deglaciation. When exposed with a spade the rock displays a pristine surface of glacial striation and polish (Figure 6.3a); left bare, this will degrade in fewer than ten years. There are many examples to be seen in local stone quarries. The Bruce Peninsula and Manitoulin Island in Lake Huron are excellent Holocene karsts precisely because wave action in early postglacial lakes removed most of the till, leaving the bedrock unshielded for the past 10 000 years.

Anticosti Island, Gulf of St Lawrence, also contains much beautiful postglacial karst. Large tracts of it have no surface streams or ponds because water passes directly underground via karren. But there are occasional patches of marl (lime-rich clay). Where no more than 25 cm in depth these have performed the same shielding role. String bogs are well developed upon them (Roberge 1979; Roberge and Ford 1983). The transition from a full marsh pond to dry holokarst karren field can occur in 2–3 m.

At Windsor, Nova Scotia, the last glaciers erased a pinnacle karst in gypsum, leaving an undulating, scoured surface indented by small doline bottoms. This surface is approximately 12 000 years in age and, although in a soft, very soluble rock, is beautifully preserved beneath 4–6 m of clay-rich till containing gypsum clasts that have

FIGURE 6.6 (a) A small stream drains into a large, permanently inundated doline in the Goose Arm glaciated karst, Newfoundland. (b) For comparison, a doline and mogote tropical karst in Puerto Rico. The scale and form of these two landscapes are quite similar but the Newfoundland karst aquifer has been clogged with deep, subglacial injecta

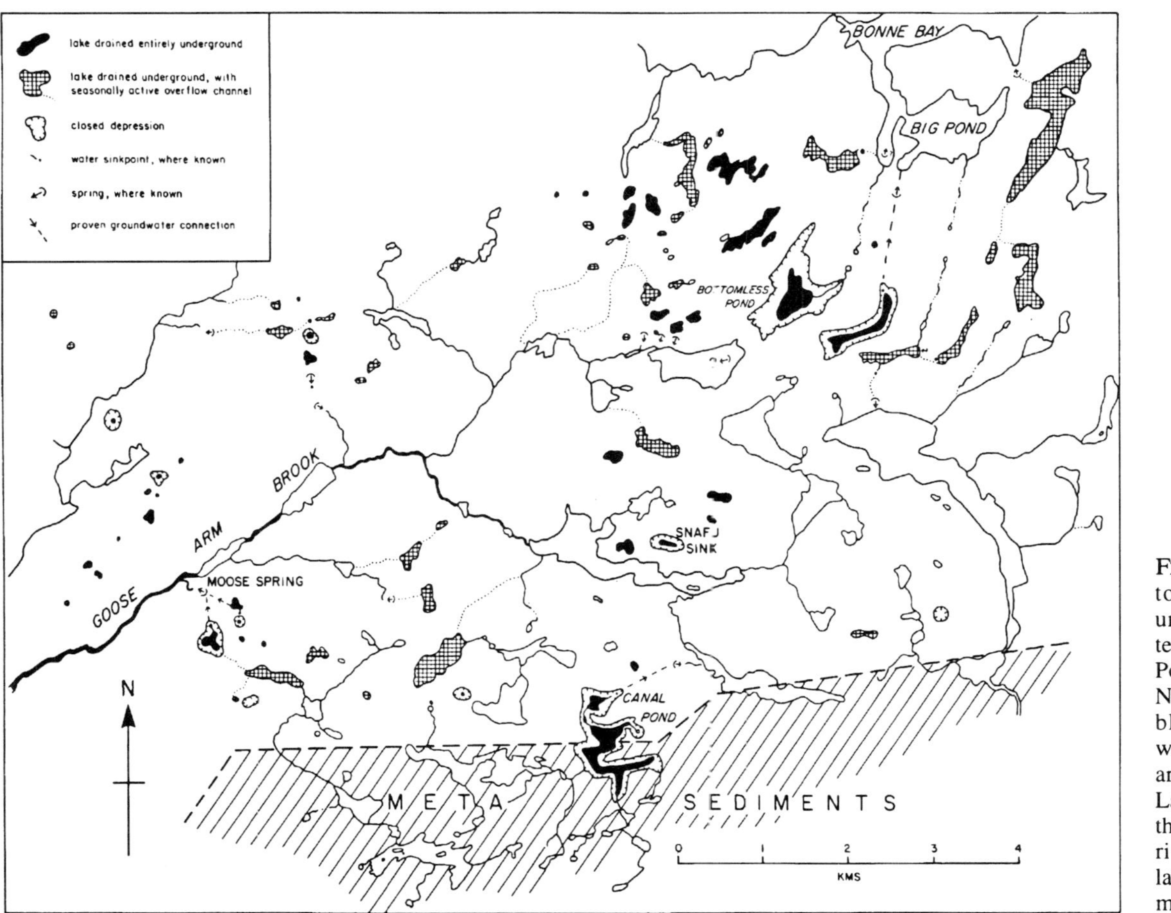

FIGURE 6.7 Schematic map to illustrate surface and underground drainage patterns in the Bonne Bay Big Pond — Goose Arm karst, Newfoundland. Lakes in black discharge all their water underground; springs are indicated, where known. Lakes in white discharge all their water through surface river channels. Chequered lakes drain underground but may also overflow during the spring thaw melt season

protected it. It can be seen around the edges of modern gypsum quarries. A few hundred metres away tidal rivers removed the till and a dense new pinnacle karst has developed during the postglacial period.

The destructive or inhibitive impacts of glaciers upon karst have been emphasized thus far in this review because they appear to have been predominant in Canada. However, much terrain beneath sheets of flowing ice suffers negligible glacial erosion because the glacier is frozen at the base (e.g. Sugden 1978). On limestone and dolomite, the consequence is that there may be good to perfect preservation of karst features as fragile as karren beneath a layer of melt-out till that was deposited as the ice wasted. This is the *sealing* effect.

The principal example lies beneath Winnipeg and its environs. It is a dense clint-and-grike pavement in dolomites that covers perhaps 3500 km^2 (Ford 1983b). This is the most extensive pavement reported anywhere. The last glaciers ripped up a few patches of it but most is preserved beneath till that is overlain by Glacial Lake Agassiz clays (Figure 6.3b). The till partly infills the top one metre of the pavement but is not deeply injected. Groundwater flows into the pavement from higher ground around the city and the Agassiz clay seals this in. As a consequence the pavement has been an excellent aquifer. It supplied Winnipeg (with many water-intensive meat packing plants) until World War I when, under emergency circumstances, the capacity of the aquifer was exceeded and it became contaminated with deeper salty waters (Render 1970). It still supplies suburban communities.

The final group of effects recognizes that glacial action may stimulate karst development, especially during general glacier recession when enormous quantities of ice that had accumulated slowly are released rapidly in the form of meltwater. Large melt streams can be directed at potential karst input points at a glacier sole or along its margins for a few tens or hundred of years, and areally integrated glacier aquifers can be superimposed upon karst aquifers in the rock. These processes of *focusing streams* and *raising hydraulic head* are most effective in rugged alpine terrains such as the Rocky Mountains. It is probably true to write that for the carbonate rocks the fastest global rates of development of small and medium-scale karst landforms, particularly of cave systems, occur in these circumstances.

The evidence for focusing is the quite widespread occurrence of relict dolines, shafts or cave entrances located in what become hydrologically improbable positions once the ice has gone. Examples include shafts in the crests of roches moutonnées (where they were evidently drilled by recurrent glacier moulin streams) or aligned along the outer edges of recessional moraines. Most impressive are sink points or entire, short karst groundwater systems (sinkhole, cave and spring) that occur high in the walls of major valleys. Evidently supraglacial margin melt streams flowed into the rock for a while and then returned to the ice.

Superimposition of a glacial aquifer onto a karst aquifer may be a powerful mechanism (Figure 6.8). It operates most effectively in glacially overdeepened depressions in the mountains; in the Canadian Rockies there are some hundreds of overdeepened cirques that are now drained through the limestone in their floors. An extreme example of superimposition is found on the Columbia Icefield, the largest remaining ice mass in these mountains: a small lake in a depression that is entirely within the ice builds up in some summers until it is emptied abruptly through its base (Muir and Ford 1985, plate 8).

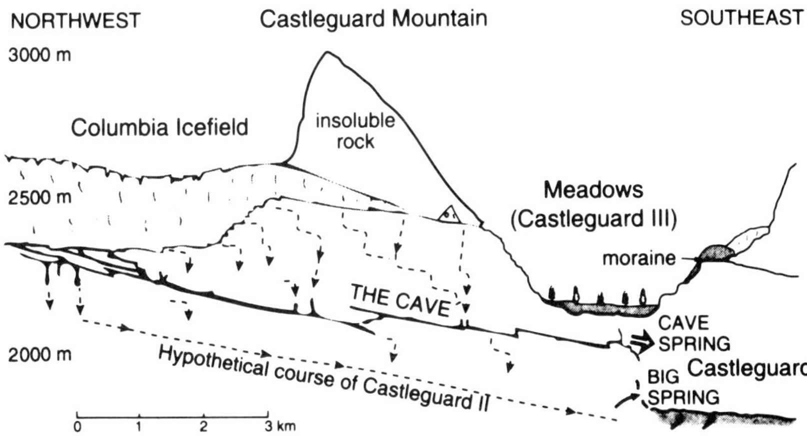

FIGURE 6.8 Schematic section through the Castleguard karst, Rocky Mountains, showing a portion of the Columbia Icefield superimposed on the karst aquifer. "Castleguard II" is a modern system of inaccessible caves discharging subglacial meltwater some hundreds of metres beheath Castleguard Cave, a drained, relict feature. "Castleguard III" denotes a second inaccessible system that underdrains alpine meadows and a part of the Saskatchewan Glacier

It appears that it discharges into the underlying karst aquifer and must be characterized as a "karst jokulhaup" (glacial lake dam burst flood).

Karst development is also stimulated where glaciers entrench major valleys because this steepens the hydraulic gradient between sink and spring by *lowering the elevation of the springline*. In the western mountains many small karst systems are unable to keep pace with this comparatively rapid lowering because their springs "hang" in the walls at some tens of metres above the floor although there are no impermeable strata to perch them there.

At one remove, glacial entrenchment has been responsible for initiating many short cave systems in lowland Canada because glacier scour of bedded strata creates many local scarps. Where postglacial river courses are established to flow across them or parallel and above them it is common to find that some or all of the water is captured into underground shortcuts. Bonnechere Caves, Ontario, are an excellent example where three successive caves developed to divert waters from a small gorge to progressively lower spring points (Ford 1962): none of the caves was longer than 150 m. Wodehouse Creek, Ontario, is an instance of parallel capture along the edge of a major glacial valley (Cowell and Ford 1975). Beaupre and Caron (1986) describe many similar examples from the Quebec karst areas.

The final special effect of glacial action upon karst rocks and systems is more speculative than the others. It is the proposition that at times of large-scale glacial recession and the accompanying isostatic rebound of the land there may be injection of part of the abundant meltwaters deep underground, i.e. to depths of 1000 m or more. Injection is associated with the crustal flexure that must occur during rebound. This is essentially a *one-time or 'slug' injection* mechanism, perhaps once per glacial cycle of ice-sheet growth and decay. It delivers a lot of water to deeply buried karst strata, inducing episodes of interstratal dissolution.

The most important evidence is that some deep bedrock collapse structures (breccia pipes) either were initiated or were rejuvenated at times of glacial recession. In most instances the collapses originate with the dissolution of salt or anhydrite that rests on limestone or dolomite and is covered by other rocks, including clays and shales that normally are impermeable unless subject to unusual stress such as flexing. The greatest examples known in Canada are found in southern Saskatchewan because the top of the salt formation there (Figure 6.2) has been quite intensively explored for potash deposits. A first instance is Howe Lake (Christiansen 1971). This is a depression where Wisconsinan (last glacial) tills display cylindrical faulting with a downthrow of 78 m. Collapse occurred immediately after the ice receded. The depression has been filling slowly with local soilwash, airborne dust etc. throughout the Holocene and is now only 10 m deep although it is 300 m in diameter. The "Saskatoon Low" (Figure 6.9) is a much greater geological feature but it has no modern topographic expression. From the data and analysis of Christiansen (1967), in one or a series of early collapses Upper Cretaceous and older shales were downfaulted at least 190 m to create one large depression measuring 25 × 40 km, plus three lesser ones on its flanks. Christiansen considered that

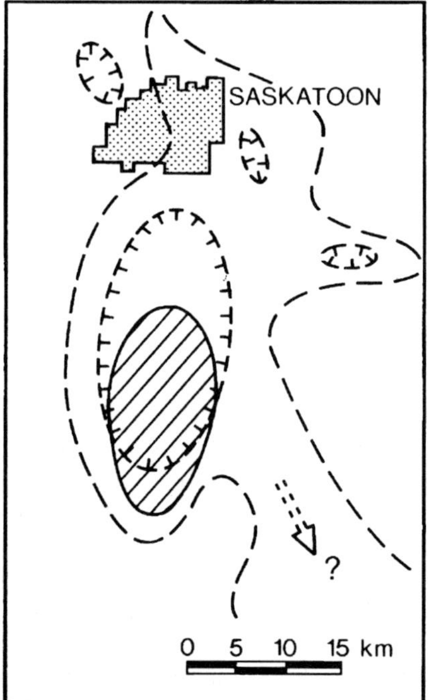

FIGURE 6.9 The "Saskatoon Low" (adapted from Christiansen 1971, with permission). Broken hachures mark location of four closed depressions in bedrock that are now concealed by mantles of glacial and glaciolacustrine debris. The cross-hatched feature is a younger collapse inside the largest of these depressions: collapse of 70 m occurred at the close of the last (Wisconsinan) glaciation

these collapses could have occurred during the Tertiary, but there is no exceptional hydrological mechanism to produce them then and no evidence to indicate that they are not Quaternary glacial in their age: I suspect that they are. The most recent collapse was contained largely within the older feature and occurred precisely at the close of the last glaciation. Tills of that age downfaulted 70 m, then the new depression was infilled entirely with proglacial lake sediments. It has been stable in postglacial time. The collapse propagated from salt with its base resting on dolomite 1100 m below the modern prairie surface.

Evidence of glacial rejuvenation of *paleokarst* (karst rendered inert by infill and/or burial beneath later consolidated strata) that is of Devonian age has been revealed in the Pine Point zinc/lead mining district south of Great Slave Lake, NWT. The ores fill paleocaves and collapse dolines in a barrier reef (Rhodes et al. 1984). Certain of the dolines now display cores of Quaternary till (or of a sequence of tills) passing down through the centres of their older breccia and sulphide fillings, indicating that there has been a renewal of vigorous groundwater circulation. Modern hydraulic gradients are too low to support such circulation so we must suspect that it occurred during the deglaciations.

This last effect emphasizes perhaps more than the others the problems encountered in dealing with karst in glaciated environments. In both the Saskatchewan salt and the Pine Point dolomites, ancient rocks and paleolandforms that had been inert for more than 250 million years have been subject to vigorous glacial karstic erosion even though they are covered by substantial depths of later, insoluble rocks.

KARST, GLACIATIONS AND PERMAFROST

The complexity and variability of glacial effects upon karst is further enhanced in the permafrozen regions of Canada. Because of the cost of access there have been comparatively few studies. The operation of karst groundwater processes and the special restraints placed upon them are poorly understood as a consequence. Figure 6.10 is a first, highly generalized, attempt to model them along a traverse poleward from the southern limits of permafrost.

Karst features may host and sustain permanent ground ice far south of the conventional limits of arctic permafrost. Where large cave chambers descend steeply from the surface they will trap dense winter air and retain it throughout the summer unless they receive direct sunlight or there are vigorous water or air drains out of the bottom. Such chambers are cold traps. Waters seeping into them become frozen and may accumulate (with seasonal layering as in normal glaciers) to depths of several tens of metres, or even hundreds of metres in some shafts. These ice bodies are termed "glaciéres" (Balch 1900). At the present time we know of only small examples in the Canadian mountains. These are conserved where mean annual external temperatures are as high as 0°C or a little greater. Marshall and Brown (1974) used oxygen isotope ratios of the ice at a cave in Crowsnest Pass to suggest that the glaciére was at least 4000 years in age. The most celebrated glaciére is Dobsinska Ice Cave in the hills of southern Slovakia; a volume of $\sim$150,000 m^3 of ice is preserved in a cold trap where the mean annual external temperature is +6°C, a very high value.

Development of karst systems does not appear to be significantly restricted in zones of

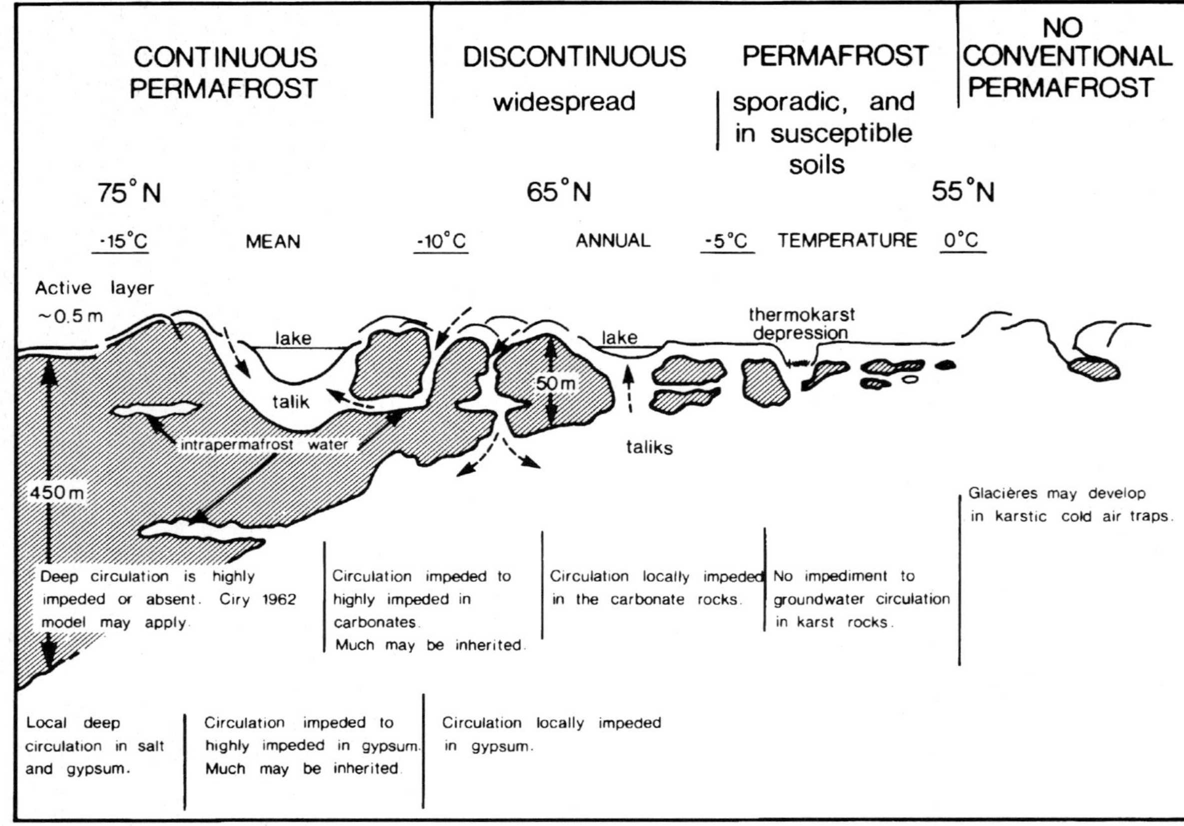

FIGURE 6.10 A general model displaying the relationships between karst and permafrost in Canada

sporadic permafrost. Permanent freezing in these regions is confined to soils such as silts which readily accrete ground ice. If these accumulate in traps such as doline bottoms ice growth might temporarily obstruct circulation but Cowell (1983) investigated a low-gradient, very low-energy karst system along the Attawapiskat River in the Hudson Bay Lowlands sporadic zone and found no permafrost restraints.

In the more severe zone of widespread but discontinuous permafrost, freezing extends into the bedrocks and, as a consequence, the situation becomes more complex. In the Mackenzie region several different studies have been made across this zone and into the margins of continuous permafrost north of it. They compose a picture of increasing restriction. The southerly (and most intensively studied) site is the Nahanni karst, at Lat. 61°N in the Mackenzie Mountains. It is a rugged terrain of deep dolines and solutional corridors (Figures 6.11 and 6.12). The mean annual temperatures fall from −5° to −7°C with increasing elevation. There is unimpeded, well-integrated karst drainage from all larger depressions to springs at the two ends of the karst belt. Between the large depressions, the highest karst is permafrozen and relict. An intermediate zone displays aperiodic impedance of the recharge. Its catchments are small, a few hectares at most. The drains of dolines there become sealed by ice or by detrital plugs that freeze. Over one or a few melt seasons water accumulates above these seals until the hydrostatic pressure is sufficient to rupture them; then they drain in the space of a few hours or days (Brook 1983; Brook and Ford 1978). An important point with respect to the Nahanni karst is that it has been glaciated, but not for the past 350 000 years or more. The modern karst

FIGURE 6.11 A small closed polje becoming inundated in the Nahanni karst, Mackenzie Mountains, NWT. The nearer cliffs seen on the left-hand side are 20–30 m in height

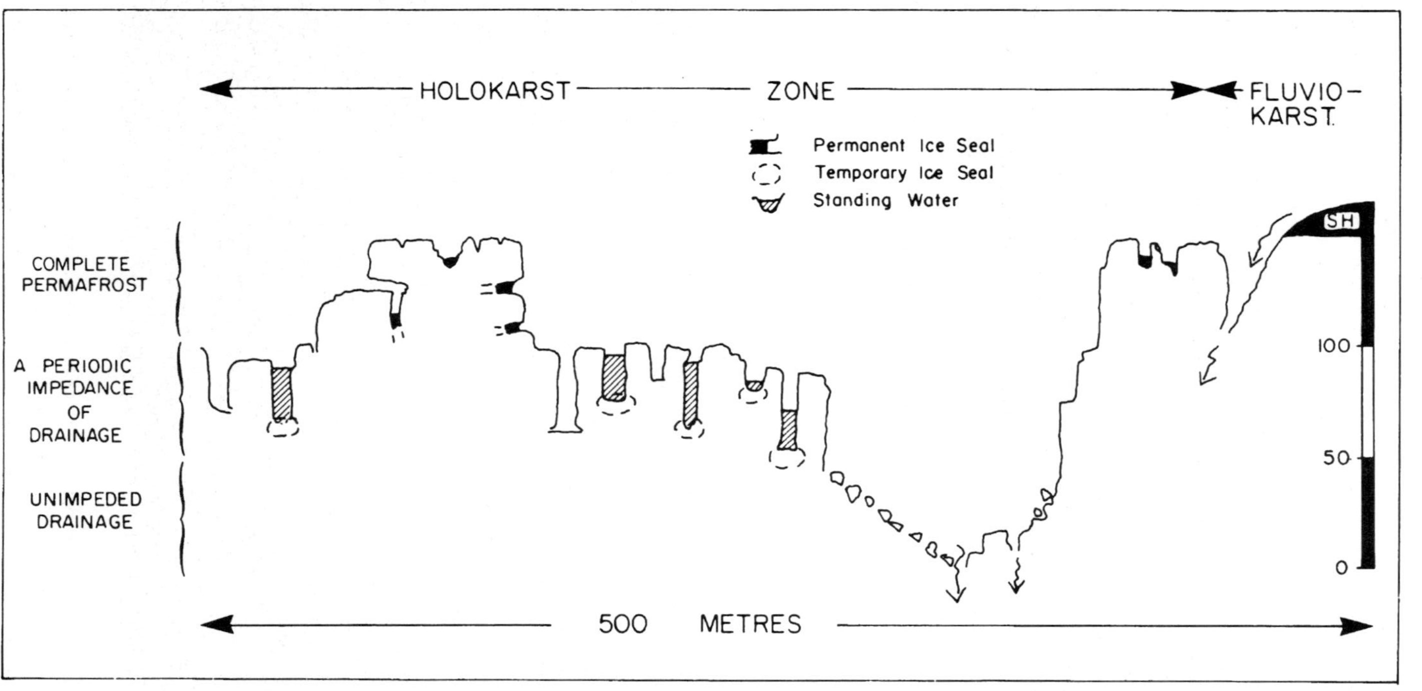

FIGURE 6.12 A model for mountain karst development in the Discontinuous Permafrost zone. The model is based upon the situation in the Nahanni North Karst, Canada

landforms developed after that glaciation and thus were generated in climatic conditions generally much more severe than those that exist in the region today.

To the northeast in the Mackenzie lowlands there are low hills and shallow plateaus around the west end of Great Bear Lake and in the Horton and Anderson river plains north of it. The gradients to sustain groundwater circulation are much reduced except along escarpment edges. The Discontinuous–Continuous Permafrost boundary bisects the region. Nevertheless, there are dolines scattered irregularly throughout it, composing a belt that is up to 125 km in width. The belt is underlain by limestones, dolomites and gypsum. All of it was glaciated during the last (Wisconsinan) glaciation and most areas are veneered with glacial deposits that mask the rocks.

Van Everdingen (1981) studied the southern half of the belt, mapping 1400 dolines, 27 larger, polje-like depressions and at least 63 substantial karst springs. Karst in limestone and dolomite alone is small and confined to escarpment edges. The doline karst develops where groundwaters can penetrate a thin cover of these rocks to gain access to the more soluble gypsum. There, the karst system has certainly been maintained as the permafrost grew back in during postglacial times; indeed, it appears to have expanded quite vigorously at sites where the hydraulic gradients are comparatively steep (Figure 6.13). However, it is probable also that the major karst systems were initiated before postglacial times, perhaps during the lateglacial time of isostatic rebound and abundant meltwater upon the thawed ground, perhaps in earlier and warmer interglacials or even in preglacial

FIGURE 6.13 A polje in the lowland karst west of Great Bear Lake, NWT. The feature is 4 km in length. It is developed in a veneer of Wisconsinan glacial and proglacial sediments resting on 30–40 m of Ordovician dolostone underlain by gypsum and anhydrite. (Photograph by Dr R.O. van Everdingen, with permission)

FIGURE 6.14 *For caption see opposite*

times (though I would doubt the latter).

Dolines become fewer north of the Discontinuous limit but they do extend for at least 200 km into the Continuous Permafrost zone at the Mackenzie lowlands (Yorath et al. 1968). There are no detailed studies but (from scanning the air photographs) the impression is that the extent of active karst groundwater circulation may be shrinking. This implies that it was most extensive during the deglaciation and has been reduced as the permafrost grew back.

Deeper into the Continuous Permafrost zone (i.e. on the arctic islands) modern karst is yet more limited. On limestone and dolomite it is mostly small (karren scale) and confined to scarp edges or to the active layer (seasonally thawed layer) on bare rock slopes. This conforms to a simple model of restricted karstification that was proposed by Ciry (1962). Lauriol and Gray (1993) describe excellent examples on Akpatok Island, Hudson Strait, which has a comparatively mild, maritime climate (Figure 6.14). Woo and Marsh (1977) report a sample of the more limited development found in the severe, high arctic conditions of Ellesmere Island.

There are evidences that normal, quite widespread and deep karst groundwater circulation and dissolution did occur beneath or at the melting margins of the last glaciers at some sites in these regions, however. There are patterns of intersecting solution corridors in limestone or dolomite at lowland sites on King William, Prince of Wales and Somerset islands and the Adelaide Peninsula (Fraser and Henoch 1959; Bird 1967) and on the rugged Borden and Brodeur peninsulas of Baffin Island. These corridors are relict and inert today; so also are dolines in gypsum on Devon Island that are very well preserved and thus seem likely to have been created by meltwater streams at the margins of the Wisconsinan glacier. Finally, in the Nanisivik zinc/lead mine on Borden Peninsula we can view permafrost from the inside. The mine is in dolostones and has an ambient temperature of −11°C. For a length of 200 m along its south wall, dolostone and ore (both Precambrian in age) are a jumble of giant blocks in a state of arrested collapse. The arresting agent is a firm cement of ground ice. Evidently waters were injected (to a depth of at least 60 m) into the bedrock beneath a valley glacier and dissolved the dolostone to induce the collapse. This process halted when the water froze following removal of the insulating ice cover (Ford 1984; Figure 6.15).

CONCLUSIONS

The principal conclusion must be that the effects of glacier action upon the karst systems and permafrozen terrains which are so widespread in Canada can be extremely varied — in fact they seem to encompass the gamut of possibilities that I can imagine. At one

---

FIGURE 6.14 (*opposite*) Postglacial limestone pavements in the temperate and permafrost zones compared. (a) Typical clint-and-grike terrain at the crest of Malham Scar (escarpment) in Yorkshire, England. This site was scoured by ice during the last (Devensian) glaciation. (b) Clint-and-grike terrain at the crest of limestone excarpments on Akpatok Island, NWT. This site is in the Continuous Permafrost zone. The surface of the pattern of intersecting grikes seen at Malham is here reduced to a thermoclastic scree (felsenmeer). However, continuing effective karstic drainage downwards beneath it is indicated by the aligned micro-depressions seen in the scree in the foreground. (Photograph by Professor Bernard Lauriol, by permission)

1. LAST INTERGLACIAL?

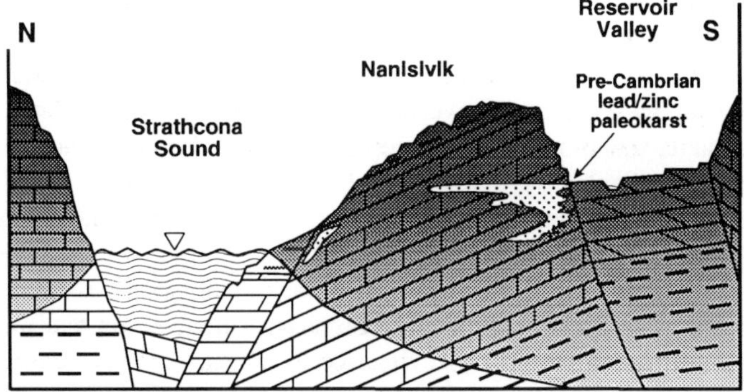

2. LAST GLACIATION

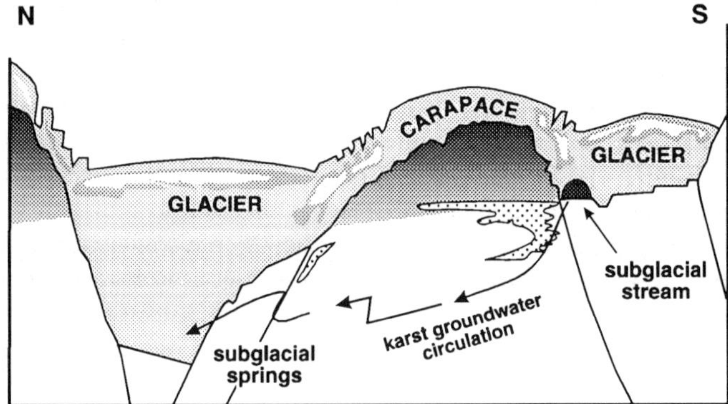

3. HOLOCENE

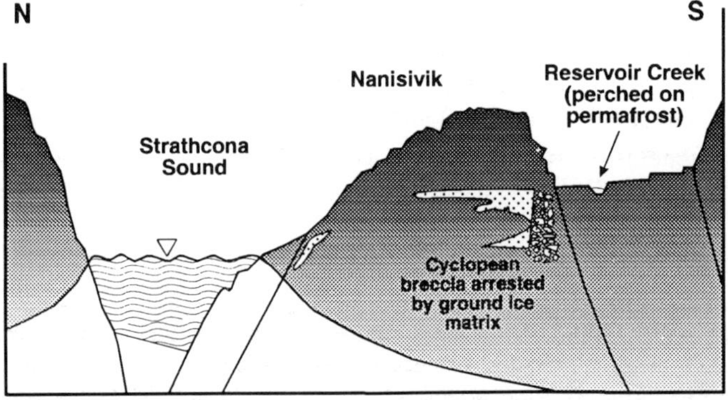

FIGURE 6.15 *For caption see opposite*

extreme there is the complete destruction of any pre-existing karst landforms, as on many limestone surfaces in Ontario and Quebec; at the other is the near-perfect preservation of the most fragile forms, the example of clint-and-grike pavement beneath Winnipeg. There is the total prohibition (during all of postglacial time) of dissolution upon those very extensive surfaces of limestone and dolomite that are shielded by lime-rich tills to be contrasted with the most rapid genesis of sinkholes and caves that can occur at the margins of melting glaciers. In the more soluble gypsum, these zones of total prohibition and of fastest genesis of karst may occur within a few tens of metres of each other, depending upon the vagaries of till deposition and its postglacial removal.

These effects constrain the validity of modern measurements of karst processes in glaciated regions. The seeming efficiency or vigour of a process may be a product of precursor glacier action. For example, some of the highest concentrations (at thermodynamic equilibrium) of dissolved Ca, Mg and CO_3 ions ever recorded in simple bicarbonate waters are to be found in our Canadian karsts, even on the purportedly less soluble dolomite. Yet ironically those particular waters are not generating any new karst at all; they gain their dissolved solids entirely from the shielding lime-rich tills because CO_2 gas, water and rock interact simultaneously there in optimally efficient (open system) three-phase reactions. We may measure contemporary processes at work at postglacial karst sites such as many pavements, but the results should not be extended to other pavements until it is established that the latter also are truly postglacial, i.e. that they lack earlier grikes which survived the glacier scour (Lundberg and Ford 1994). Very often it is not feasible to eliminate the possibility of such an inheritance. Modern process studies are not applicable when considering those large-scale features of interstratal dissolution that here have been attributed to slug injections of meltwater during regional glacier recession.

In the permafrost regions geomorphic relationships become yet more complex. There is little impediment to the operation of karst processes at southerly and high hydraulic gradient sites such as the Nahanni karstland, but further north it appears that deglacial times may have been the times of most significant karst genesis and thus that the modern karst systems are inherited to quite an extent. There is a close analogy with the paraglacial phases and landforms of clastic sedimentology and glacial geomorphology (Church and Ryder 1972). In the arctic islands the most vigorous karst development probably took place beneath glaciers. It could not be sustained into postglacial times and thus the karst features became permafrozen relicts, as they are at Nanisivik.

This is not to "knock" the studies of present-day geomorphic processes that have been such a major part of academic geomorphology in the English-speaking world for the past

---

FIGURE 6.15 (*opposite*) Schematic cross-sections through the Nanisivik lead/zinc ore body (Baffin Island, NWT) to model the development of a mega-breccia arrested by ground ice. 1. Before brecciation (the last Interglacial or earlier). The terrain is ice-free and permafrozen to depths >100 m except beneath Strathcona Sound, an arm of the ocean. 2. The brecciation event (the last (Wisconsinan) glaciation?). Permafrost largely or entirely thaws beneath a cover of glacier ice. Subglacial meltwater is diverted underground through the Nanisivik karst (Proterozoic dolostones) to Strathcona Sound. Dissolution induces the collapse of the southern part of the ore body along a frontage of ~200 m. 3. Postglacial. Permafrost conditions return. Brecciation is halted by the growth of ground ice

25 years (Bremer, this volume, Chapter 2). They will always be valid provided that the researcher can appreciate and specify the limits to that validity. It is the unique feature of the karst geomorphic system that its channels are located underground where glaciers can rarely impact them directly. The consequences are the wide range of effects and interactions discussed here that continue to be influential in the postglacial landscape and environment.

For fun I teach planetology, the morphology and processes of other planets and moons in the solar system so far as these are known. This teaches humility. There are certainly some general laws of landform development that can be applied to most of these bodies. None evade gravitational effects. Particular processes are spatially dominant on many, crude bombardment being the chief among them. But each of the larger bodies displays at least a few extensive patches of morphology that does not occur on any other. The 'weird terrain' of Mercury, the Venusian coronae, Martian chaoses, the ice mountains of Ganymede, mega-scarps of Miranda and 'cantaloupe terrain' of Triton, are all good examples from the surfaces that have been imaged thus far. The morphology of the earth is much richer in its diversity than any of these other bodies. It is to be expected that it will display general causal complexity and may contain local assemblages of landforms that are truly unique. In the context of the Great Debate on geomorphology in this volume (Chapter 2), and the discussion that it has stimulated elsewhere (Richards 1994), this is to adopt a position of 'pragmatism' or 'scientific realism'. It recognizes that geomorphologists are historians of natural landscapes as much as they are scientists. The radically varying impacts of glacial processes that can be detected in the karst regions of Canada are an interesting example of the complexity.

REFERENCES

Balch, E.S. 1900. *Glacières or Freezing Caverns.* Lane & Scott, Philadelphia.

Beaupre, M. and Caron D. 1986. *Decouvrez le Quebec souterrain.* Presses de l'Univ. du Quebec, 254 pp.

Bird, J.B. 1967. *The Physiography of Arctic Canada, with Special Reference to the Area South of Parry Channel.* Johns Hopkins University Press, 336 pp.

Brook, G.A. 1983. Application of LANDSAT imagery to flood studies in the remote Nahanni karst, Northwest Territories, Canada. *Journal of Hydrology*, **61**, 305–324.

Brook, G.A. and Ford, D.C. 1978. The origin of labyrinth and tower karst and the climatic conditions necessary for their development. *Nature*, **273**, 493–496.

Brown, M.C. 1972. *Karst Hydrology of the Lower Maligne Basin, Jasper, Alberta.* Cave Studies 13, Cave Research Associates.

Christiansen, E.A. 1967. Collapse structures near Saskatoon, Saskatchewan, Canada. *Canadian Journal of Earth Sciences*, **4**, pp. 757–767.

Christiansen, E.A. 1971. Geology of the Crater Lake collapse structure in southeastern Saskatchewan. *Canadian Journal of Earth Sciences*, **8**(12), pp. 1505–1513.

Church, M. and Ryder, J.M. 1972. Paraglacial sedimentation: A consideration of fluvial processes conditioned by glaciation. *Bulletin, Geological Society of America*, **83**(3), pp. 3059–3072.

Ciry, R. 1962. Le role du froid dans la speleogenese. *Spelunca Memoires*, **2**(4), pp. 29–34.

Cowell, D.W. 1983. Karst hydrogeology within a subarctic peatland. *Journal of Hydrology*, **61**, pp. 169–175.

Cowell, D.W. and Ford, D.C. 1975. The Wodehouse Creek karst, Grey Co., Ontario. *Canadian Geographer*, **XIX**(3), pp. 196–205.

Ford, D.C. 1962. The Bonnechere Caves, Renfrew Co., Ontario. *Canadian Geographer*, **V**(3), pp. 22–25.

Ford, D.C. 1979. A review of alpine karst in the southern Rocky Mountains of Canada. *National Speleological Society of America Bulletin*, **41**, pp. 53–65.
Ford, D.C. (ed.) 1983a. Castleguard Cave and Karst, Columbia Icefield Area, Rocky Mountains of Canada. *Arctic and Alpine Research*, **15**(4), 425–554.
Ford, D.C. 1983b. Karstic interpretation of the Winnipeg Aquifer. *Journal of Hydrology*, **61**, pp. 177–180.
Ford, D.C. 1984. Karst groundwater activity and landform genesis in modern permafrost regions of Canada. In LaFleur, R.G. (ed.), *Groundwater as a Geomorphic Agent*, Allen & Unwin, London, pp. 340–350.
Ford, D.C., Schwarcz, H.P., Drake, J.J., Gascoyne, M., Harmon, R.S. and Latham, A.G. 1981. Estimates of the age of the existing relief within the southern Rocky Mountains of Canada. *Arctic and Alpine Research*, **13**(1), pp. 1–10.
Fraser, J.K. and Henoch, W.S. 1959. Notes on the glaciation of King William Island and Adelaide Peninsula, NWT. *Department of Mines and Technical Surveys, Geographical Branch Paper*, **22**, 54 pp.
Karolyi, M. 1978. Karst development in Ordovician carbonates: Western Platform of Newfoundland. M.Sc. thesis, Geography, McMaster University, 171 pp.
Karolyi, M.S. and Ford, D.C. 1983. The Goose Arm Karst, Newfoundland, Canada. *Journal of Hydrology*, **61**, pp. 181–185.
Lauriol, B. and Gray, J.T. 1993. Drainage karstique en Milieu de Pergelisol: le Cas de L'Ile d'Akpatok, T.N.O., Canada. *Permafrost & Periglacial Processes*, **1**, pp. 129–144.
Lundberg, J. and Ford, D.C. 1994. Canadian landform examples — 28 dissolutional pavements. *Canadian Geographer*, **38**(3), 271–75.
Marshall, P. and Brown, M.C. 1974. Ice in Coulthard Cave, Alberta. *Canadian Journal of Earth Sciences*, **11**(4), pp. 510–518.
Muir, R.D. and Ford, D.C. 1985. *Castleguard.* National Parks Centennial, Supply and Services Canada, 120 pp.
Render, F.W. 1970. Geohydrology of the metropolitan Winnipeg area as related to groundwater supply and construction. *Canadian Geotechnical Journal*, **7**(3), 243–274.
Rhodes, D., Lantos, E.A., Lantos, J.A., Webb, R.J. and Owens, C. 1984. Pine Point ore bodies and their relationship to structure, dolomitization and karstification of the Middle Devonian barrier complex. *Economic Geology*, **70**, pp. 991–1055.
Richards, K., 1994. 'Real' geomorphology revisited. *Earth Surface Processes and Landforms*, **19**, 277–281.
Roberge, J. 1979. Geomorphologie du Karst de la Haute-Saumons, Ile d'Anticosti, Quebec. M.Sc. thesis, McMaster University, 217 pp.
Roberge, J. and Ford, D.C. 1983. The Upper Salmon River karst, Anticosti Island, Quebec, Canada. *Journal of Hydrology*, **61**, pp. 159–162.
Rose, L. and Vincent, P. 1986. Some aspects of the morphology of grikes: a mixture model approach. In Paterson, K. and Sweeting, M.M. (eds), *New Directions in Karst.* GeoBooks, Norwich, pp. 497–514.
Saunderson, H.C. 1977. The sliding bed facies in esker sands and gravels: a criterion for full-pipe (tunnel) flow. *Sedimentology*, **24**, 623–638.
Smart, C.C. 1988. Quantitative tracing of the Maligne karst system, Alberta, Canada. *Journal of Hydrology*, **98**, pp. 185–204.
Sugden, D.A. 1978. Glacial erosion by the Laurentide ice sheet. *Journal of Glaciology*, **20**(83), pp. 367–392.
Sweeting, M.M. 1972. *Karst Landforms.* Macmillan, London, 362 pp.
van Everdingen, R.O. 1981. Morphology, hydrology and hydrochemistry of karst in permafrost near Great Bear Lake, Northwest Territories. *National Hydrological Research Institute of Canada. Paper*, **11**.
Woo, M-K. and Marsh, P. 1977. Effect of vegetation on limestone solution in a small high Arctic basin. *Canadian Journal of Earth Sciences*, **14**(4), 571–581.
Yorath, C.J., Balkwill, H.R. and Klassen, R.W. 1968. Geology of the eastern part of the Northern Interior and Arctic Coastal Plains. *Geological Survey of Canada, Paper*, 68–27.

7

A Meltwater Model for Laurentide Subglacial Landscapes

JOHN SHAW

Department of Geography, University of Alberta, Edmonton, Canada

ABSTRACT

It is proposed that many subglacial landforms result from erosion and deposition by catastrophic outburst floods. Actualistic and deterministic approaches to glacial geomorphology are described and assessed. The generation of models by fantasy is introduced as an important step in the evolution of a meltwater model for subglacial landscape evolution. Subsequent inductive testing of this model involves comparison of observed and expected characteristics of landforms.

Erosional marks in bedrock (s-forms) give the most compelling evidence for meltwater sheet flows. Spindle flutes, sichelwannen, muschelbrüche, and potholes are best interpreted as products of erosion by vortices in turbulent flows. Remnant ridges behind obstacles (rat tails) and associated hairpin scours are well explained by the erosional effects of horseshoe vortices generated by bluff obstacles in turbulent flow. The absence of striations or only weakly developed striation in many of these s-forms and percussion marks caused by the collapse of cavitation bubbles in potholes and downflow from protruberances also point to meltwater, rather than glacial, erosion. Maps of s-forms, including furrows and large-scale fluting, around the northeastern shore of Georgian Bay and in the Kingston area, Ontario, together with estimated paleocurrent directions, point to their formation under a broad sheet flow with discharges of at least 10^7 m^3/s. Regional-scale outburst floods are implied.

The outburst flood hypothesis began with a form analogy between drumlins and marks eroded by separated, turbulent flows. Drumlins fitting this analogy are interpreted, on the evidence of form and internal structure, to be infills of inverted erosional marks scoured in the ice bed by subglacial meltwater. Their shapes (parabolic, spindle and transverse asymmetrical) are quite different from the tapered shapes of classical drumlins and are, therefore, renamed, Livingstones. Some Rogens are also considered to be cavity fills. Other, more classical drumlins are interpreted as part of a continuum from remnant ridges within hairpin scours to isolated streamlined, classical forms. Members of this continuum are named Beverleys. Sculpting of Beverleys by horseshoe vortices and accelerating fluid passing over and between residual ridges is supported by Reynolds Number estimates for the flows. A striking form similarity, over several orders of magnitude in scale, between Beverleys and yardangs also points to erosion by turbulent flow. Like the bedrock erosional marks, drumlins are considered to indicate enormous, regional-scale outburst floods.

Geomorphology Sans Frontières. Edited by S. Brian McCann and Derek C. Ford.

The Finger Lakes, New York State, and Lake Mistassini, Quebec, are tentatively interpreted as products of meltwater erosion. Continental-scale mapping of flowlines and the relationships between drumlins, scour zones, striations, eskers, glaciofluvial "interlobate moraines", and arcuate, glaciofluvial end moraines suggest simultaneous formation of drumlins beneath much of the Laurentide ice sheet. A large supraglacial lake, connecting suddenly to the subglacial hydrologic system, is postulated in a conjectural model for meltwater storage and release that may have formed drumlins, Rogens and scoured bedrock.

Extraglacial effects of floods on fluvial systems, marine sedimentation, ocean currents and climate are briefly discussed. Their effect on sea level and ice-sheet stability may explain collapses of the Laurentide ice sheet through Hudson Strait and the formation of Heinrich layers in the North Atlantic Ocean.

INTRODUCTION

In this article, I present the hypothesis that meltwater processes may have dominated the evolution of deglacial landscapes during the waning stages of large parts of the Pleistocene ice sheets. While eskers and tunnel valleys are acknowledged to be meltwater landforms and some of the Earth's largest end moraines — the Salpausselkä, Cree Lake and Sakami moraines — are clearly glaciofluvial, there remains a tendency to treat meltwater processes as secondary to ice. Water at the ice bed is also realized to be essential for sliding. Furthermore, the importance of porewater pressure to the strength and deformability of glacial substrates is recognized in recent attempts to model fast glacier flow, ice-sheet behaviour, and subglacial bed formation (Boulton et al. 1985; Alley et al. 1987; MacAyeal 1993). Nevertheless, the major subglacial bedforms are generally attributed to subglacial deformation and ice-moulding, or the accretion of lodgement till (cf. Menzies and Rose 1989).

In recent years, a small group of Canadian geomorphologists has suggested that regional-scale, outburst floods best explain extensive fields of subglacial bedforms (cf. Shaw et al. 1989; Kor et al. 1991). If this alternative interpretation is correct, classical landforms such as drumlins, flutings, Rogen moraines and erosional marks or s-forms (more commonly termed p-forms) are of glaciofluvial rather than glacial origin.

To date, this possibility has simply added to the puzzlement about these enigmatic landforms because there is still a multitude of plausible explanations. Despite the profusion of theory, there can only be one true origin for an individual drumlin, or Rogen moraine ridge, or fluting, but, without directly observing their formation, we cannot determine this origin with certainty. Understanding these landforms is made even more difficult if, for example, the features we call drumlins were formed in different ways. This was probably the case; I think that, in the past, landforms that differ morphologically and genetically have been lumped together as drumlins. To give an anecdotal illustration, a prominent glacial geologist once told me that alleging certain ridges in front of a modern glacier were drumlins was like calling a horse a dog because they both have four legs. Clearly, this confused terminology creates its own problems.

Besides the risk of mistakenly lumping together different features, thornier problems arise because we cannot be sure that morphologically similar landforms are genetically related. Karcz (1973) had this in mind when he spoke poetically of the lethal lure of geometric similarity. However, for geomorphology to be practical, we must assume that linguoid ripples on a river bed have a common genesis, as do barchan dunes in a sand sea, and ridges in Rogen moraine.

By direct observation, we have no difficulty in attributing fluvial ripples and eolian dunes to the work of water and wind, respectively. We have no equivalent observations for some subglacial landforms. Though there are good reasons for concluding that they were formed subglacially, the fact remains that their formation has not yet been observed. Because glacial and fluvial processes are both at work beneath ice sheets, we cannot *assume* that one or the other was the formative agent for a given landform. It appears, in fact, that there is no concensus on the agent of formation for, say, drumlins. Little wonder then that the processes of formation are heatedly debated.

It is, therefore, timely to return to fundamentals, to question assumptions and to carefully dissect competing hypotheses on subglacial bedforms. A comprehensive review of hypotheses on subglacial landscape genesis is beyond the scope of this essay. As there are numerous extensive reviews of more conventional approaches to drumlin formation (e.g. Boulton 1987; Menzies and Rose 1989), an overview of the meltwater alternative may provide balance and an opportunity to compare competing hypotheses. Hence, this is mainly an account of a meltwater hypothesis for subglacial bedforms: drumlins, fluting, Rogen moraine and s-forms.

GEOMORPHOLOGICAL SCIENCE

It is difficult to discuss the meltwater model without briefly considering the methodologies by which subglacial bedforms are investigated. This is necessary because different approaches give rise to widely differing hypotheses — a lack of convergence and concensus that seriously compromises application of theory in glacial geomorphology.

The *actualistic* approach has been most influential in recent years. In its most conservative form, actualism holds that only processes that are observed directly give acceptable explanations. Thus, in this approach, observations of subglacial processes beneath the margins of modern glaciers are the mainspring for process interpretations of modern and ancient landforms. Detailed kinematics and dynamics can be discovered by directly observing and measuring subglacial processes. As well, there is much to be learned from the quasi-actualistic study of landforms where they emerge from the forefront of modern glaciers (Figure 7.1).

As an example of actualism, detailed measurement of deformation rates and patterns in sediment beneath Icelandic glaciers (e.g. Boulton 1979) leads many to infer that bed deformation explains Pleistocene subglacial bedforms (cf. Menzies 1989). The actualistic linking of ice streams and subglacial deformation (cf. Alley et al. 1987) and subglacial deformation and drumlins (Boulton 1987) are the chief justification for linking drumlins and ice streams (Dyke and Morris 1988; Boyce and Eyles 1991). By this reasoning, ice streams are inferred to have formed drumlins by deforming the glacial substrate. This view prevails in recent literature despite observations that the ice/bed interface for Ice Stream B, a fast-flowing stream of ice in West Antarctica, is very flat, and fluting at the deforming-till/bedrock contact would not appear in the landscape upon deglaciation (Alley et al. 1987). Alley et al. (1987) considered fluting to be enigmatic. Moreover, this view is accepted even though I am unaware of any studies in which deformation explains

FIGURE 7.1 Woodworth Glacier and forefront, Tasnunana, Valley, Alaska (Bradford Washburn). The flutings behind boulders and the prominent fan with a feeding esker allow confident reconstruction of subglacial processes occurring when the present forefront was beneath the glacier

drumlin form, assemblages, landform associations, landscape setting, and the sedimentary evidence better than the meltwater hypothesis presented here. Actualism must be applied with caution to landforms for which there are no truly actualistic observations.

Experiments in flumes or with materials testing equipment are also actualistic. They serve to provide models and tests for imaginative hypotheses emerging from field observations on sediment and landforms of past glaciers.

A *deterministic* approach, whereby theory on the physics of glaciers is used to argue for or against hypotheses of subglacial bed formation, could provide powerful tests of their validity — a hypothesis that contravenes fundamental principles of physics is unlikely to attract many supporters. Were such physical theory available, it might explain the variety of drumlin forms, their locations, and the timing of their formation. Unfortunately, glacier physics has not yet been so successful. At present, explanations of glacial bedforms from physics are no better than the geomorphological assumptions upon which they are based. Thus, a deterministic theory based on meltwater erosion is not very useful if drumlins actually form by till accretion.

A final approach to geomorphological explanation involves *fantasy* and *inductive testing*. Images of how landforms are created are generated by fantasy (I find it hard to understand how landforms, for which there are no actualistic examples, can be understood without such creative thought), with the validity of the imagined processes tested by comparing observed landform characteristics with those predicted by the fantasized model. Analogy probably inspires the fantasy. Thus, the meltwater hypothesis for drumlins came about because I was surprised one afternoon to realize the great similarity between drumlins and flutes on turbidite beds. Denis Johnson, a colleague in human geography with little experience in geomorphology or sedimentology and no prejudice on drumlin formation, and I compared photographs of sole marks in Dzulynski and Walton (1965) with air photographs of the Livingstone Lake drumlins. He had no difficulty in matching drumlins and flutes. My children are equally good at matching the shapes of drumlins to fluvial and eolian erosional marks.

The form similarity almost immediately prompted in my mind an imagined environment and mechanism of drumlin formation — subglacial meltwater sheets eroded cavities into the overlying ice bed and drumlins are positive counterparts of these inverted erosional marks (flutes). Many years later I realized that regional-scale floods beneath the Laurentide ice sheet are implicit in this model of drumlin formation. These floods may explain Rogen moraines and s-forms in bedrock as well as drumlins.

Similarly, fantasy inspired the comet hypothesis for drumlins (Figure 7.2). Obviously, such a scientific hypothesis must go beyond fantasy; it must proceed to careful testing of the images it inspires. Fantastic or outrageous hypotheses may be retained only if critical tests do not call for their rejection. Thus, to uphold the comet explanation for drumlins, material in the drumlins must be similar to sediment and rock in and around James Bay

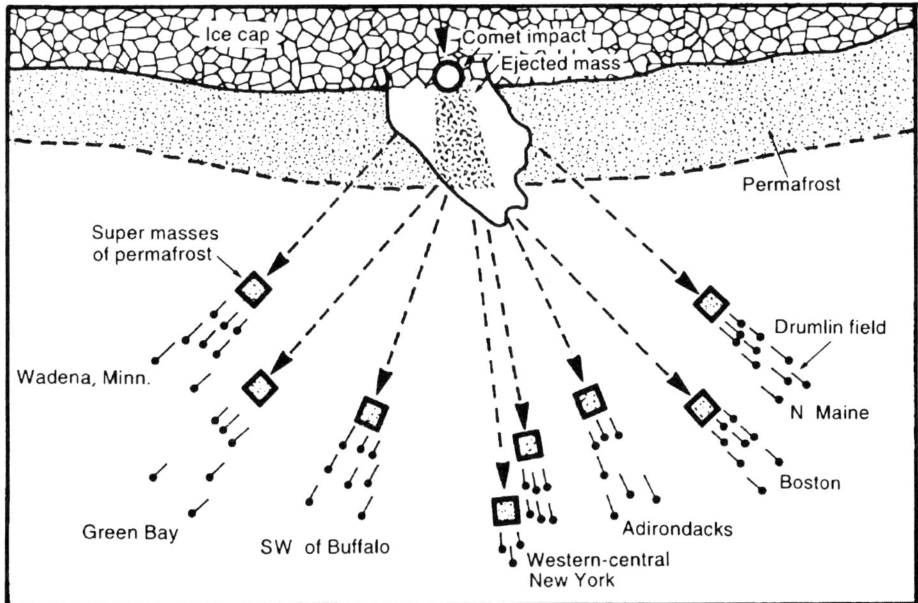

FIGURE 7.2 Drumlin formation resulting from a comet impact. Ejected masses of permafrozen ground disintegrate on landing producing drumlins as splash marks (after C.A. Davis, unpublished)

(Figure 7.2). In addition, only by chance would the orientation of drumlins in this hypothesis correspond to ice-flow directions indicated by striae and dispersion paths in their vicinity. Since clasts in drumlins are mainly locally derived and drumlin axes are usually aligned with the trend of local ice-flow directional indicators, the evidence contradicts the comet hypothesis.

The meltwater hypothesis for subglacial bedforms is not rejected by similar inductive tests. Rather, the hypothesis is found to explain an increasingly wide variety of otherwise enigmatic bedforms and to combine them successfully in a single model. This model and the evidence used to test it are at the heart of this essay. Although *actualistic* and *deterministic* elements are used in this testing, *inductive reasoning* plays the critical part in leading *fantasy* towards sound theory.

THE MELTWATER MODEL

It is best to present the model for meltwater bedforms first and then proceed to inductive tests by which observations are compared to expectations derived from the model. To visualize the postulated subglacial environment, imagine a continental ice sheet in which a broad, flow-parallel zone of the glacier is separated from its bed by a fast-flowing sheet of meltwater (Figure 7.3). This meltwater erodes and deposits at the bed — the locations of these processes may shift with time. The model shows areas of dominantly eroded landscape where bedrock is widely exposed at the surface. Drumlins may be purely erosional remnants of pre-existing sediment or bedrock, or they may be constructional hills, formed as fills of cavities carved upwards into the ice. Rogen ridges are depicted in Figure 7.3 as purely depositional, cavity-fill forms, though some may well be mainly erosional (Fisher and Shaw 1992).

It is crucial that the meltwater sheet had an erodible upper surface (Figure 7.3). Non-uniformity of the flow, resulting from variable geometry between the ice bed and the substrate, was the fundamental reason for land formation by differential erosion and deposition. Deposition or reduced rates of erosion in cavities and increased rates of erosion at flow constriction resulted in a distinctive landscape (Figure 7.3).

Shaw and Kvill (1984) first studied the so-called cavity-fill drumlins in the Livingstone Lake field, northern Saskatchewan to test the meltwater model. Subsequent research on bedrock erosional marks provides a simpler and more convincing test because these purely erosional forms are remarkably similar to other forms known to have been produced by turbulent fluids and it is hard to explain bedforms in granite and gneiss by the competing subglacial deformation model (Shaw 1988a; Kor et al. 1991). Hence, bedrock erosional marks are discussed together with erosional forms in sediment. Then, cavity-fill drumlins and Rogen moraine are described and interpreted to complete the inductive test of the meltwater hypothesis.

EROSIONAL MARKS – S-FORMS

Introduction

Hall (1815) first described eroded bedrock surfaces in and around Edinburgh, Scotland,

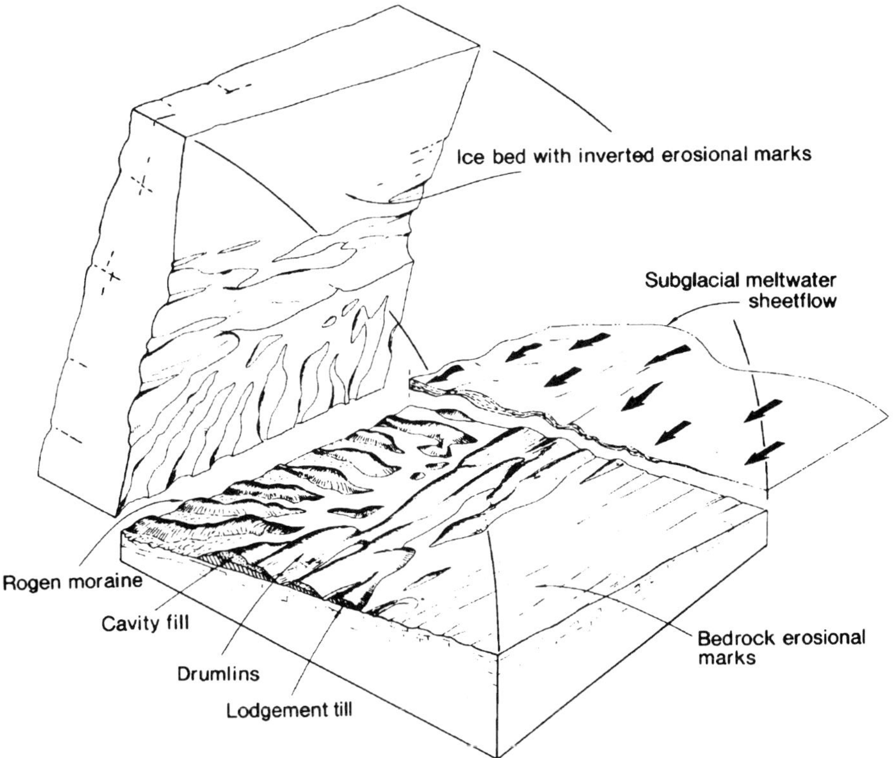

FIGURE 7.3 A model for subglacial landforms produced by meltwater outburst floods. Bedrock erosional marks and some drumlins are depicted as purely erosional features. Cavity-fill drumlins and Rogen moraines are depicted as depositional forms infilling erosional marks in the ice bed. Drawing by John Glew

with long ridges and intervening erosional troughs. He compared these landforms with features in wind-eroded snow to suggest that the rock had been sculpted by giant tidal waves. Since then this interpretation, based on form analogy, has changed only in that meltwater is now considered the eroding agent rather than tidal currents. Ljungner (1930) described a number of new forms from his field studies in Sweden and first proposed a meltwater origin. Dahl (1965) and his students mapped erosional marks in Norway and noted their tendency to be best formed on slopes facing into the flow. He also suggested that the discontinuity of marks on rock surfaces indicated alternating reaches of subglacial and englacial meltwater flow.

Allen (1971) opened the modern era of research on erosional marks by verifying experimentally that a number of erosional mechanisms produce similar forms. He demonstrated convincingly that these erosional forms represent interaction between fluid and bed and that separated flow structures are essential for erosional mark formation. Allen (1982) included subglacial forms in a general classification of erosional marks. Kor et al. (1991) relied heavily on Allen's classification in their comparison of erosional marks scoured in crystalline bedrock around the north shores of Georgian Bay, Canada, and marks observed or inferred to result from erosion by turbulent flow. Shaw (1988a) and Sharpe and Shaw (1989) also argued for meltwater erosion of hairpin grooves and rat-tail ridges.

Other researchers have suggested direct glacial erosion of most subglacial marks (Smith 1948; Gjessing 1965; Boulton 1974; Goldthwait 1979; Laverdière and Guimont 1980; Laverdière et al. 1985). Laverdière and Guimont (1980) and Laverdière et al. (1985) illustrated several new forms. Preferential abrasion by streaming debris-rich ice is the favoured erosional mechanism. I have reasoned against this hypothesis elsewhere (Shaw 1994) and mention it here simply to point out that there are published alternatives to the meltwater interpretation of erosional marks.

FIGURE 7.4 Open-spindle flutes, French River, Georgian Bay, Ontario. Flow toward viewer. The open-spindle flutes are left of centre, their sharply pointed, leading ends to the left of the knife. Comma forms and sichelwannen lie upflow of the spindle flutes

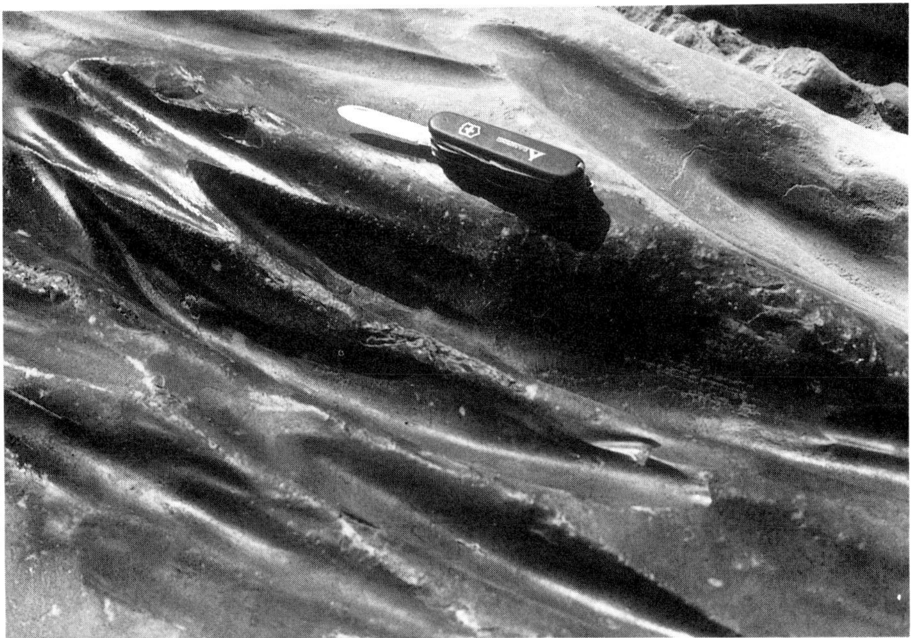

FIGURE 7.5 Closed-spindle flutes, Wilton Creek, Ontario. Flow from bottom right to top left. These flutes show characteristic asymmetry; to an observer looking downstream, the right sides of the flutes are relatively steep and the right rims relatively straight

Form and process

Spindle flutes are spindle-shaped with a long axis parallel to the flow direction. Whereas closed-spindle flutes have sharp, continuous rims, open flutes have sharp rims on the upflow side and merge imperceptably with the surrounding rock surface in the downstream direction (Figures 7.4 and 7.5). Spindle flutes are usually asymmetrical about the flow-parallel axis: one rim is more curved than the other and the cross-sectional slope is lower on the side with the more curved rim (Figure 7.5). The leading (furthest upstream) end of spindle flutes is usually sharper than the trailing end. Shaw (1988a) illustrated spindle flutes with curved axes. Spindle flutes in nature and those produced in flumes commonly occur together with other forms.

Allen (1971, 1982) attributed spindle flutes to erosion by paired vortices rotating in opposite directions and eroding by bringing high-velocity, turbulent fluid to the bed. The absence of internal morphology indicating double vortices led Shaw (1988a) and Kor et al. (1991) to conclude that spindle forms may also be formed by a single vortex impinging on the bed (Figure 7.6). These forms are found on turbidity current beds (Allen 1982), on ventifacted bedrock and wind-eroded snow surfaces (Shaw et al. 1989), on the walls of gorges eroded by fast-flowing rivers (Maxson 1940; Selby, 1985) and are common at a variety of scales in subglacially eroded bedrock (Murray 1988; Shaw 1988a; Kor et al. 1991). Since the subglacial forms are commonly only partially striated or completely unstriated and glaciers are not expected to produce impinging vortices, a glacial origin for spindle flutes is unlikely. By contrast, vortices are highly probable in

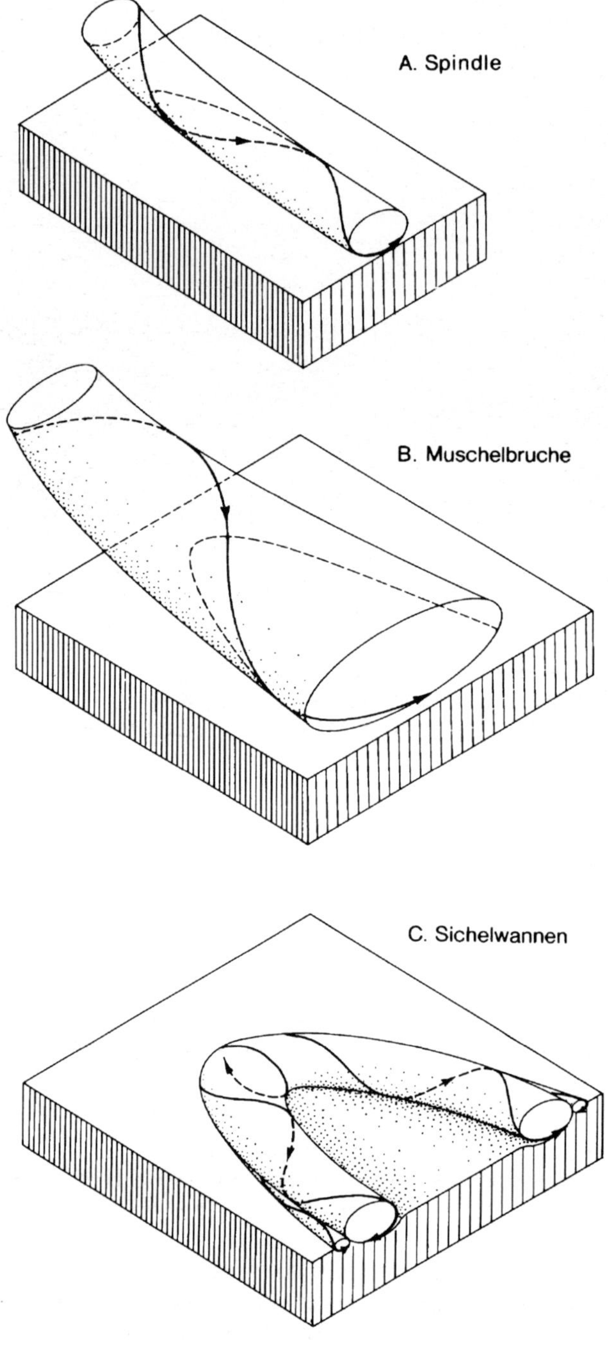

FIGURE 7.6 *For caption see opposite*

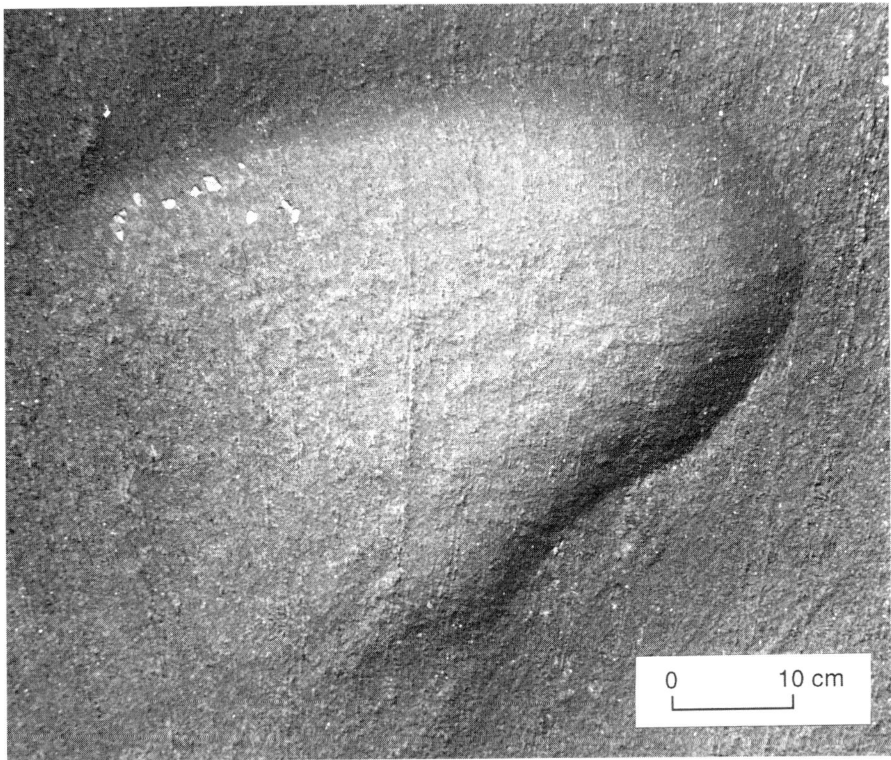

FIGURE 7.7 Muschelbrüch in granite, French River, Georgian Bay, Ontario. Flow from top right

fast-flowing meltwater and hydraulic erosion of spindle flutes is physically consistent with the turbidity current, eolian and fluvial origins of similar forms.

Muschelbrüche, first named by Ljungner (1930), are mussel-shaped depressions, with a larger width to length ratio than spindle flutes (Figures 7.6 and 7.7). They are relatively shallow, at most a few tens of centimetres deep. Like open-spindle flutes, they have sharp upflow rims and merge downflow with the surrounding rock surface. They are distinguished from sichelwannen (see below) by their simplicity of form: a longitudinal section shows asymmetry with steeper slopes below the upflow rim descending to the deepest point and very gentle slopes rising downflow towards the surrounding bed. There are no secondary ridges or troughs within muschelbrüche.

Forms like muschelbrüche were produced in flume experiments with flow confined beneath a plexiglass plate (Figure 7.8). They also occur as flute marks on turbidite beds

---

FIGURE 7.6 (*opposite*) Inferred relationships between form and flow structure for spindle flutes, muschelbrüche and sichelwannen. (A) Spindle flutes form as a result of vortex impingement at a low angle causing long, shallow erosional marks. (B) Muschelbrüche are formed by higher-angle vortex impingement causing flow to spread on contact, resulting in broad, shallow erosional marks. (C) Sichelwannen are formed when flow separation beyond the rim produces a roller eddy and vortices. Sichelwannen are self-perpetuating because they set up the eroding flow structures

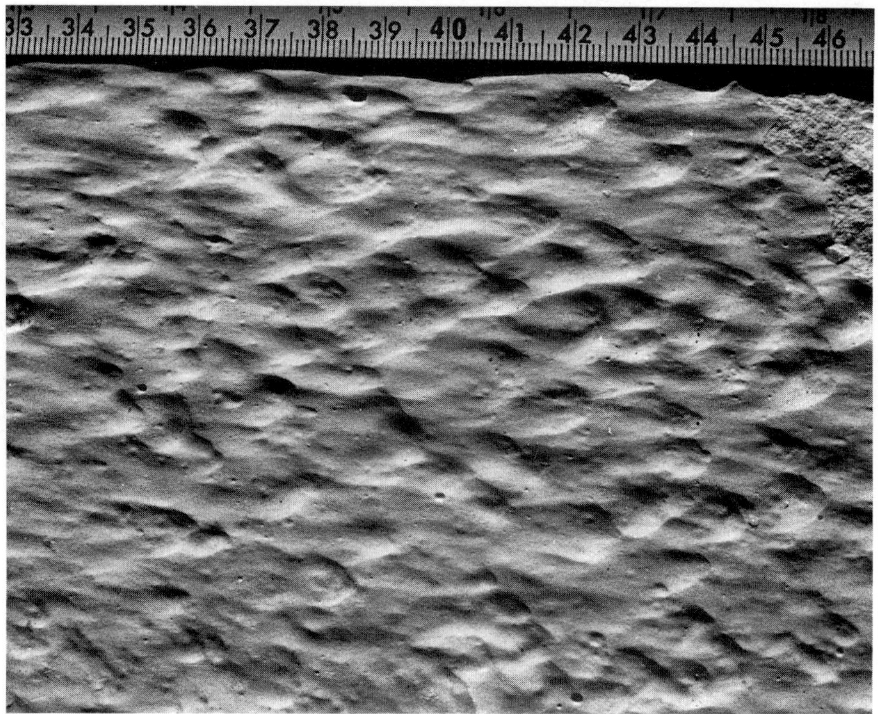

FIGURE 7.8 Muschelbrüche eroded in plaster of Paris by slightly acidic water in a flume. Flow right to left. Scale in cm

and are seen where wind has enhanced snow or ice sublimation. Like spindle forms, they are a product of erosion by turbulent fluids and may result from vortex impingement (Figure 7.6) (Kor et al. 1991). The extremely shallow incipient forms of muschelbrüche (in some cases they appear as little more than a polishing of the rock surface) counters any possibility of flow separation related to the form itself. It is this, more than any other genetic characteristic, that differentiates muschelbrüche and the more complex sichelwannen described next.

Sichelwannen are sickle-shaped erosional depressions with sharp upstream rims (Figures 7.6 and 7.9). In their interior, a prominent trough or furrow wraps around a medial ridge. Lateral furrows commonly flank the main furrow. Sichelwannen are morphologically identical to classical flutes (Allen 1982) and resemble erosional forms on cave walls (Ford and Williams 1989). They are also easily produced in flume experiments where plaster of Paris is dissolved by aggressive, turbulent water (Figure 7.10). Comma forms (Shaw et al. 1989) are closely related to sichelwannen but only one

FIGURE 7.9 Sichelwannen in gneiss, French River, Georgian Bay, Ontario. Flow from lower left

FIGURE 7.10 Sichelwannen, in the area of the plastic cubes, eroded in plaster of Paris by slightly acidic water. Note the lateral furrows. Flow right to left. Muschelbrüche are seen in the upflow part of the photograph and open-spindle flutes to the top left. Scale in cm

FIGURE 7.11 Comma-shaped s-form in granite, Hålö, Sweden. Flow away from viewer. This asymmetrical form may be considered half of a sichelwanne, only one vortex extended downflow

arm extends downstream (Figure 7.11).

The flow structure for sichelwannen is closely tied to the form itself (Figure 7.6). Flow separation downstream from the rim sets up a pair of vortices which erode the main furrow. Secondary vortices drawing energy from the primary vortices and rotating in the opposite sense erode lateral furrows. Indeed, flow structure and form are so closely related it is most improbable that sichelwannen are produced by agencies other than turbulent separated fluids. In a subglacial environment, they could only have been created by meltwater. For comma forms, only one vortex appears to have eroded the bed.

Potholes are erosional features generally accepted as resulting from vortex flow in high-velocity fluid. Vortices producing a pothole are fixed by the hole itself and their

FIGURE 7.12 Giant pothole cut into a steep rockwall. Georgian Bay, Ontario

axes are at a high angle to the bed. The interiors of many potholes carry spiral marks eroded by these vortices. Potholes associated with subglacial bedforms are commonly of large dimensions and are found along rock walls (Figure 7.12).

It is possible that many features identified as kettleholes in glaciated landscapes are actually potholes. An example is given from northern Saskatchewan where large, circular depressions are absent in drumlinized parts of the landscape (Figure 7.13). These depressions are common in large channels that dissect the drumlinized zone. Because these channels contain subglacial eskers, they are probably tunnel channels (Shaw 1983).

The channels are eroded into an older drumlinized surface; the drumlin-forming flow was clearly wider than the channelized flows. Wall effects with the creation of kolks — vortices with axes at a high angle to the bed — are to be expected in such channels, but not in the sheet flow that produced the drumlins. This sequence corresponds closely to the succession (flows characterized by longitudinal vortices to those characterized by kolks) invoked by Baker (1978) to explain the morphological details of the Channeled Scabland, Washington State. Note that the conditions under which drumlins formed are considered later and, at this stage, the meltwater model for drumlin formation is simply assumed (Figure 7.3). At the time of channel flow, ice had settled back to the bed in the areas where drumlins are preserved. In this way, potholes were formed within channels, but not on the higher, drumlinized interchannel surfaces.

Gagnon (1988) described potholes cut into mainly quartzitic bedrock of the Colline Blanche, a small hill in northern Quebec. This streamlined hill has deep potholes, with axes approximately orthogonal to the ground surface, distributed irregularly over its

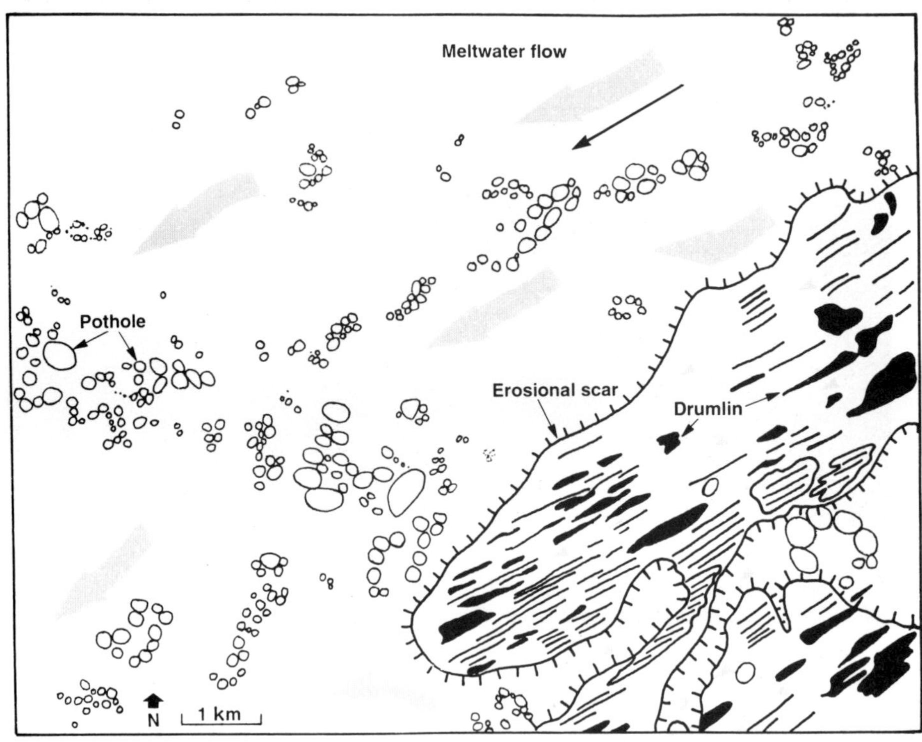

FIGURE 7.13 Potholes in broad tunnel valleys. Elevated drumlinized ground contains few potholes. Cluff Lake, northern Saskatchewan. Note that potholes are aligned in rows parallel to flow or are clustered

proximal end (Figure 7.14). The largest pothole has a diameter of 5·5 m and a depth of 9·8 m. These potholes indicate that the 45 m-high hill was submerged by a broad meltwater flow of the type illustrated in the meltwater landscape model (Figure 7.3). Potholes are generally acknowledged to form as a result of a jet spiralling downwards along the walls and rising vertically along the pothole axis (Alexander 1932; Allen 1982).

Percussion marks within the potholes and downstream from protruberances on the bedrock exposed near the crest of Colline Blanche are identical to cavitation scars on propellers and fans (Figure 7.15) (Hammitt 1980). Some percussion marks within caves are arranged in rings 5–10 cm in diameter (Figure 7.16). These are interpreted to result from cavitation in the core of small-diameter vortices embedded in the general vortical flow within the pothole. As cavitation bubbles moved away from the vortex core their collapse created the percussion mark rings.

Fairly well-constrained conditions are required for cavitation, namely that the Cavitation Number $k=2\ (p_o-p_v)/\rho v^2$ is about unity (Hammitt 1980). (p_o is water pressure, p_v is vapour pressure, ρ is fluid density and v is fluid velocity). Of course, in the presumed subglacial environment, the meltwater pressure at the bed results from the combined pressure of the overlying column of water and the overlying ice. Locally, this pressure may be reduced as a result of the flow dynamics of the meltwater. As well, flow

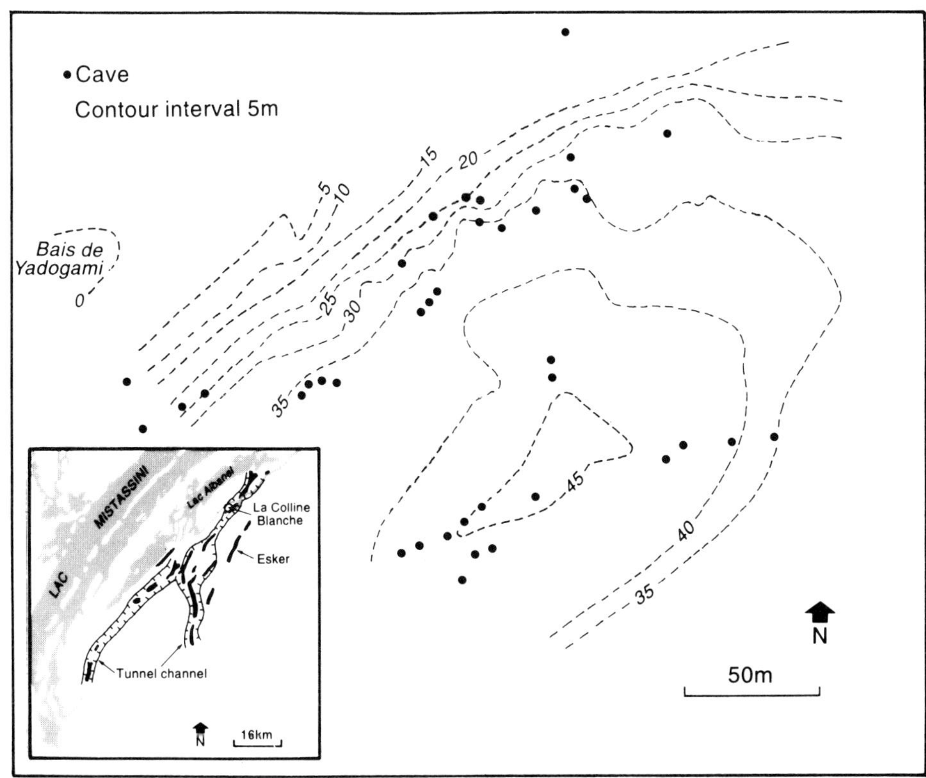

FIGURE 7.14 Colline Blanche, Quebec, showing locations of caves (potholes) and spatial relationships with major geomorphological features (inset). After Gagnon (1988)

velocities in vortices may become much larger than the mean flow, just as extraordinary wind speeds are reached in tornadoes. Consequently, taking a meltwater temperature of 0°C, and a conservative water depth of 15 m we may estimate the water velocity which would cause cavitation with changing thicknesses of ice (Figure 7.17). In the absence of local high flow velocities and low pressures, a velocity of about 60 m/s would have been required for cavitation, assuming an absolute minimum ice thickness of about 1000 m. However, much lower mean velocities in the main flow could generate local low pressures in vortices and to the lee of protruberances which would have caused cavitation.

It is obviously of interest to know whether such velocities are truly extreme or are close to other velocity estimates for catastrophic floods. Mean velocities of 30 m/s were estimated by Baker (1978) for the Washington Scabland floods and Wingfield (1990) estimated velocities of 50 m/s for flows which are inferred to have cut deep scours in the North Sea. Thus, although only extremely high velocities could have produced subglacial cavitation bubbles, similar velocities are estimated for meltwater floods elsewhere.

Large-scale fluting in bedrock and surficial deposits includes erosional grooves with intervening remnant ridges. The grooves are commonly referred to as glacial grooves, a term that implies a glacial origin for large-scale fluting. They are said to be products of enhanced glacial abrasion by streaming ice with high debris concentration (cf. Smith

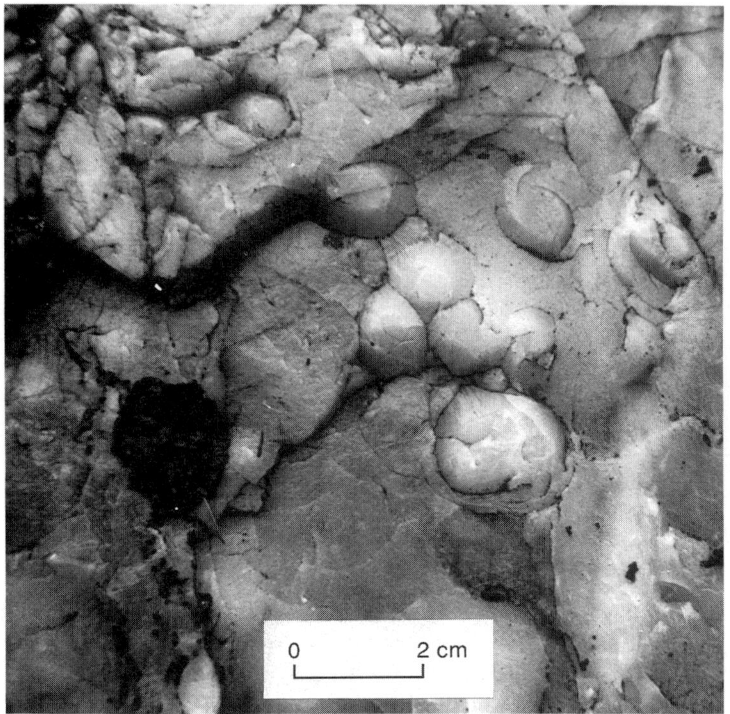

FIGURE 7.15 Percussion cones in quartzite formed by cavitation bubble collapse, Colline Blanche, Quebec. Cones are defined by concentric fractures around fluid impact points

1948). Yet, grooves in the Kingston area, Ontario, are commonly unstriated or are only lightly striated compared to the adjacent bedrock surfaces (Figure 7.18) (Shaw 1988a). Like s-forms, the Kingston area grooves have sharp rims. They are easily mapped from air photographs and indicate coherent regional flow patterns (Shaw and Gilbert 1990). They are particularly well developed where flow has been forced over an upflow-facing escarpment; the land surface at the escarpment crest is corrugated in flow-parallel alternating grooves and remnant ridges (Figure 7.19) (Shaw 1988a; Shaw and Gilbert 1990; Tinkler and Stenson 1992). These ridges are commonly tapered, forming rat tails or rock drumlins.

It is highly unlikely that unstriated grooves were formed by glacial abrasion. By contrast, a hydraulic origin is quite plausible since the Channeled Scablands floods scoured similar grooves beyond the ice margin, where glacial abrasion could not have been responsible for their formation (Baker 1978). Furthermore, the rims are sharp like those of other inferred meltwater erosional marks. Hence, the characteristics of the fluting favours a meltwater origin and the Scabland grooves support this conclusion by providing a well-documented example of hydraulically formed fluting.

Recent experimental and computational results indicate that vortices are generated upstream of escarpments facing into the flow and that these vortices are strengthened as they are swept over the escarpment rim (Wakarani 1993). The vortices have regular transverse wavelengths that decrease with increasing Reynolds Number. High bed shear

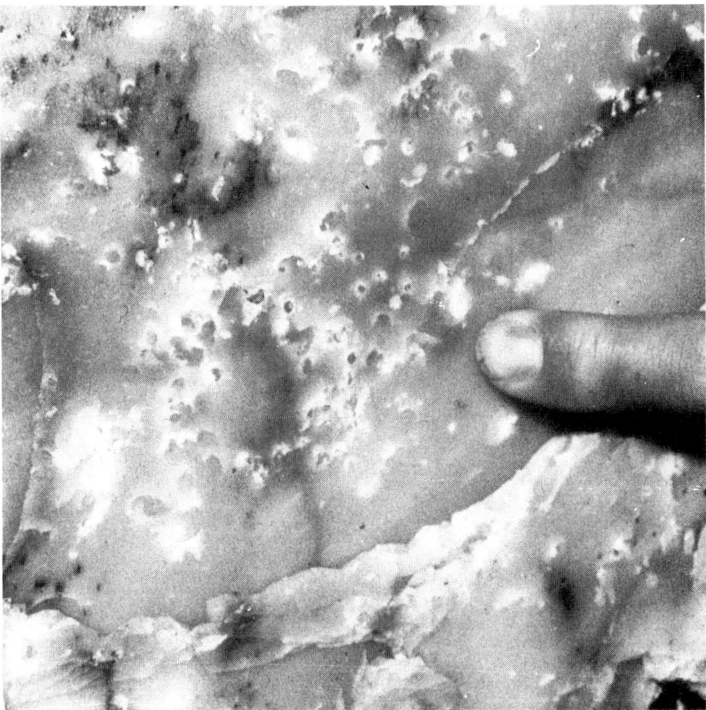

FIGURE 7.16 Percussion mark rings (e.g. left of finger) thought to be caused by bubble collapse around the core of a vortex that touched down briefly on the bed. These rings are found within the Colline Blanche potholes

stresses created by such vortices best explain the grooves above escarpments. Thus, experimental and theoretical results support inference based on form, surface texture and flow over an escarpment to suggest a broad flow overtopping prominent escarpments.

The escarpments are also sculpted into prominent flow noses and are breached by large reentrant valleys (Straw 1968; Rice 1977; Shaw and Gilbert 1990). In the past, reentrants were viewed as glacial erosional features, now they are equally well interpreted as meltwater erosional features, the scale of which conforms to the meltwater model.

Fluting is also formed in surficial deposits downflow from valley sides (Figure 7.20) (Mollard and Janes 1983). Diapiric internal structures within flutings at Athabasca, Alberta, are truncated by an erosional surface with a coarse gravel lag (Figure 7.21) and the grooves between flutings show regular, concave cross-profiles. The fluting ridges were probably formed by the erosional event that truncated the diapiric breccia. The ridges are simply remnants left upstanding when the grooves were eroded. In the same way that vortices are inferred to have formed in flow over upstream-facing escarpments, flow across valleys appears to have developed local vortices downstream from the valley rim on the downflow side of the valley. These vortices eroded the ground surface into large-scale flutings. While this reconstruction accords with a meltwater origin for the flutings, it is difficult to reconcile with known glacial processes. For example, small-scale fluting in front of modern glaciers results from constructional processes behind boulders

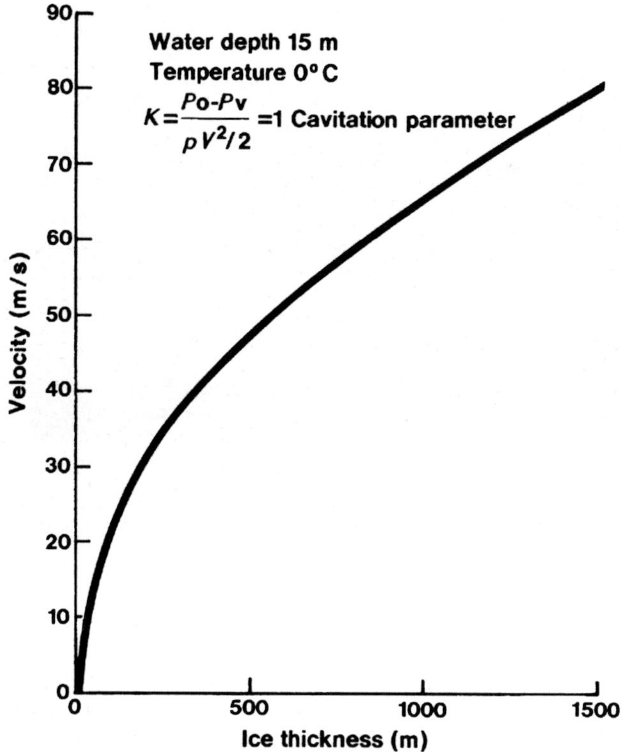

FIGURE 7.17 Water flow velocity and ice thickness for cavitation in subglacial meltwater

(cf. Dyson 1952). These processes cannot explain the preferential formation of the large-scale flutings at upstream-facing steps, nor do they explain the grooves, the boulder lag, and the truncated structures. As well, there are no obvious features, up to 20 m high, that might have acted as obstacles behind which the ridges were constructed. Finally and conclusively, constructional processes cannot explain fluting in bedrock that has the same form and landscape setting as fluting in surficial sediment.

Patterns and assemblages

To this point I have described mainly individual forms and their genesis. It is equally important to see these forms as part of assemblages or fields of bedforms with distinctive patterns. Bedforms are commonly arranged *en echelon* — sequential forms are offset in the flow-parallel direction. This arrangement, which is shown by barchan dunes, dunes in rivers and estuaries and by sole marks on turbidites (Allen 1982), is commonly shown by muschelbrüche and sichelwannen (Figure 7.22). In the case of sichelwannen, it appears as if energy for vortices that create a downstream form is, in part, drawn from the fast flow emerging from one arm of the upstream sichelwannen. This offset relationship gives rise to a very characteristic remnant rock form, blunt at the upstream end and sharply tapering downstream with concave sides in map view (Figure 7.23a). An equivalent, asymmetrical

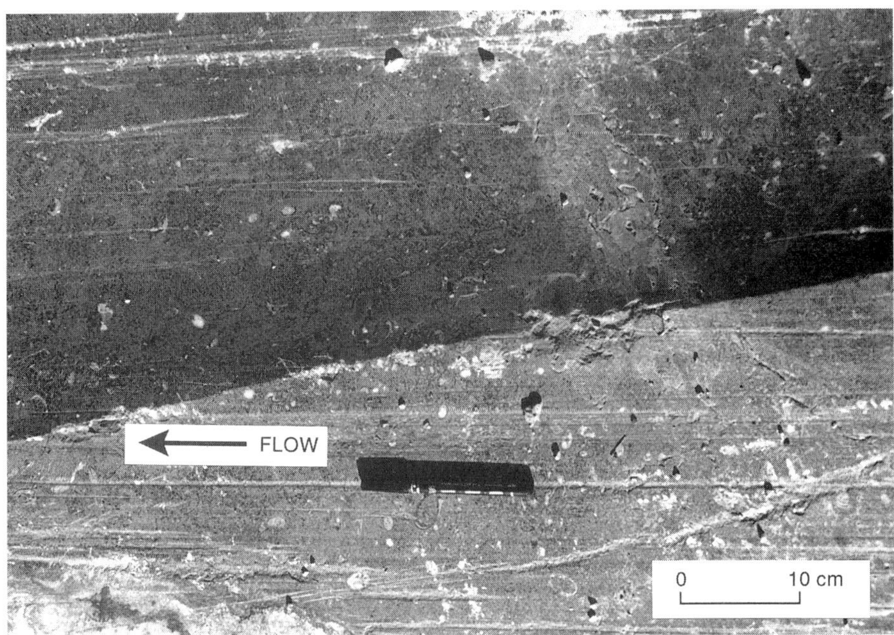

FIGURE 7.18 Fluting in carbonate bedrock, Wilton Creek, Ontario. The fluting floor (top part of the photograph) shows much lighter striation than the adjacent rock surface (bottom of photograph). To the immediate lee of the rim the fluting is unstriated

FIGURE 7.19 Fluting downflow from an escarpment in carbonate bedrock, Wilton Creek, Ontario. Note the tapered bedrock residuals or rat tails (light areas) and the prominent escarpment noses

FIGURE 7.20 Fluting in surficial deposits, Athabasca, Alberta. Flow to south (bottom). Note that the fluting starts at the downflow rim of the valley (tunnel channel) and is absent to the north of the river

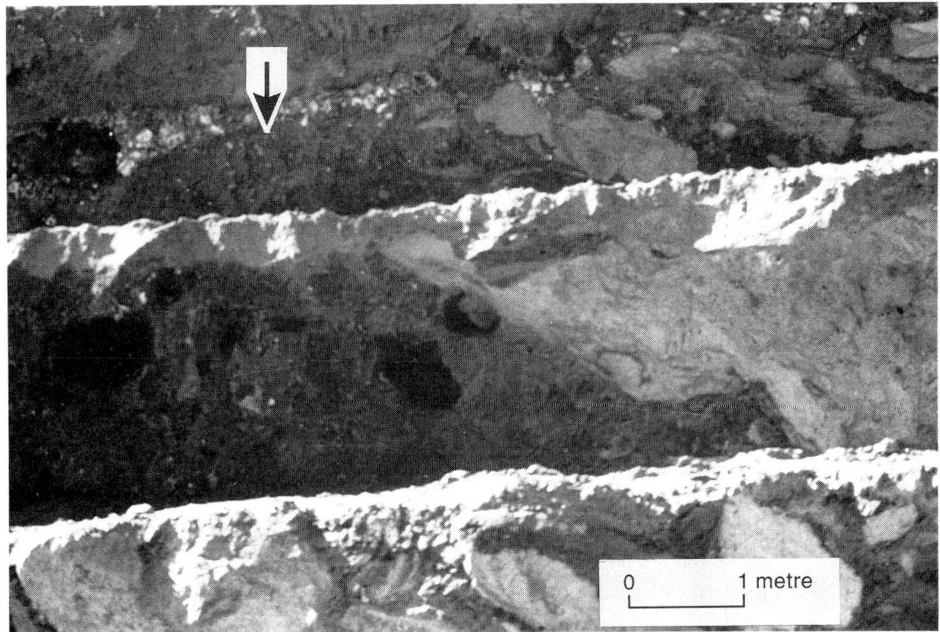

FIGURE 7.21 Brecciated sand blocks (light colour) in diapiric diamicton (dark colour) within a fluting ridge (Figure. 20). The diapiric structures were truncated during fluting formation and a gravel lag (arrow) rests on the erosional surface. Sediment overlying the gravel lag is mainly postglacial colluvium

"mutton-chop" residual ridge results where only one sichelwannen truncates an upstream medial ridge (Figure 7.23b). Residual ridges are short and have high width/length ratios where sichelwannen have short streamwise wavelengths. They are more elongate where the streamwise wavelengths are long. Sichelwannen with relatively short wavelengths are commonly clustered on proximal reverse slopes of bedrock highs, those with longer wavelengths are found on distal, downstream-facing slopes.

Furrows are distinctive linear, compound forms, generally up to a few tens of metres in width and extending several hundred metres parallel to the flow. The sides and beds of the furrows invariably carry smaller-scale s-forms (Figure 7.24) and it is this, together with the obvious scale difference, that differentiates them from the simpler grooves or flutes. Where flow was at a high angle to structural features such as bedding, jointing or foliation, furrows cross-cut the structure. Streamlined residuals with proximal crescentic scours or sichelwannen are commonly found between furrows. Smaller-scale s-forms within furrows and adjacent to them indicate that the furrows lie parallel to the eroding flow.

Regional flow

Although s-forms have been widely described and variously interpreted, Murray (1988) presented the first regional-scale map of these erosional forms. Her inferred flow

FIGURE 7.22 *En echelon* sichelwannen in gneiss, French River, Georgian Bay, Ontario. Flow away from viewer

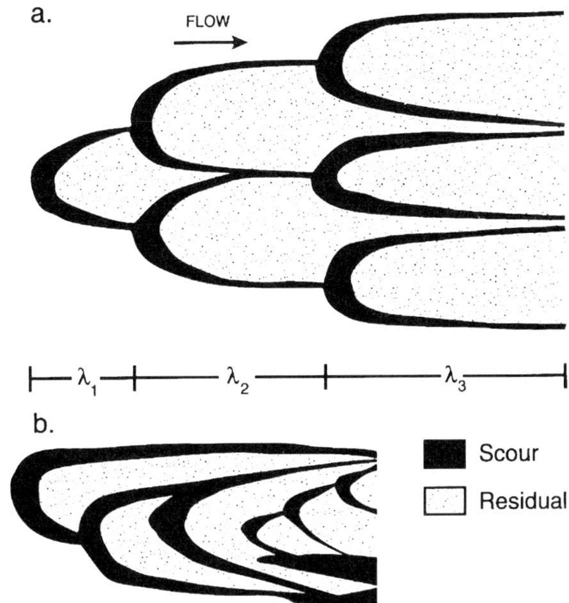

FIGURE 7.23 Forms of residual remnants resulting from erosional marks of varying wavelength (λ) and pattern

directions from s-forms in the Kingston area, Ontario, correspond closely to ice-flow direction determined from striations. Flutes formed at the rims of tunnel valleys indicate that the meltwater flows which produced the s-forms also submerged the valleys (Figures 7.19 and 7.25) (Shaw 1988a). The estimated meltwater discharges (10^6 m^3/s) can only be explained by catastrophic drainage events (Shaw 1988a).

Kor et al. (1991) also concluded that regional-scale flows created the s-forms they mapped in the French River area, Georgian Bay, Ontario (Figure 7.26). Furrow trends were mapped from air photographs and flow directions were recorded directly from s-forms. Sculpted bedrock is found almost continuously from Killarney to Pointe-au-Baril (Figure 7.26). Extensive p-forms at Whitefish Falls east of Killarney and fluting in carbonate bedrock on Manitoulin Island (Chapman and Putnam 1966) may also be part of this field. The paleoflow was unidirectional over the whole field (Figure 7.26) and aggregate rose diagrams for s-forms and glacial abrasion forms show tight clustering (Figure 7.27). The s-forms indicate paleoflows that climbed over hills and striations are superimposed on the meltwater forms. The only explanation for this is that meltwater flowed subglacially under hydrostatic pressure and striations were formed after the flood event when the glacier recoupled with the bed.

Tunnel valleys and eskers are generally sinuous and in cases even anabranched and tortuous (Figure 7.25); there was considerable variation of flow direction about a mean trend. Flow directions estimated from stratification or erosional marks could range up to 90° from the conduit trend (Brennand 1993) and, if there was conduit avulsion or flow expansion at the end of a conduit, the range of paleocurrent direction estimates could be as much as 270° (Gorrell and Shaw 1991). Thus, a wide range of flow direction is

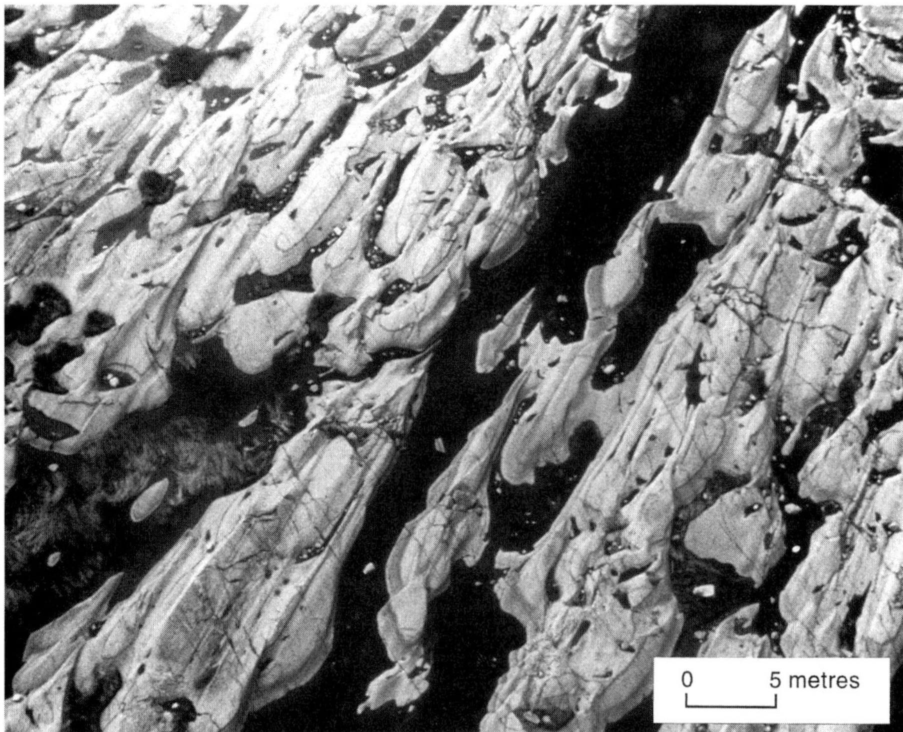

FIGURE 7.24 Water-filled furrow trending obliquely from bottom to top. Note numerous sichelwannen and a variety of residual remnants resulting from s-forms. Flow towards top right

expected for subglacial conduit flow. It is inconceivable that the tightly clustered, unidirectional paleoflow direction indicators for the broad region around the northeastern shores of Georgian Bay (Figures 7.26 and 7.27) could have originated piecemeal in narrow conduits. As well, conduit flow over a broad area and varying in position over time is expected to produce interconduit areas where fluvial erosion or deposition are absent. The continuous scour zone, together with very tightly clustered unidirectional flow estimates (Figure 7.27), argue against such piecemeal formation of the erosional marks in shifting subglacial conduits. Rather, the evidence strongly favours the interpretation that s-forms were carved into the bed of a meltwater flow at least as wide as the mapped extent of the erosional marks transverse to flow — about 70 km.

Flow magnitude

By use of the continuity equation $Q=wdv$ (Q is discharge m^3/s, w is flow width m, d is flow depth m, and v is flow velocity m/s), it is a simple matter to estimate the discharge of the regional-scale flood responsible for the Georgian Bay s-forms. Kor et al. (1991) estimated a minimum depth of 10 m to submerge rock drumlins and other upstanding areas. They also measured diameters up to 2 m for rounded, lag boulders with percussion marks. Minimum flow velocities to transport such boulders are in the range 5 to 10 m/s.

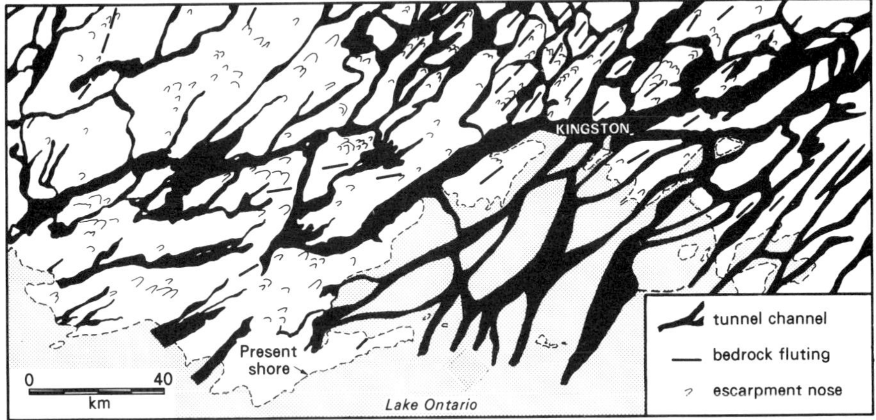

FIGURE 7.25 Anabranched, tunnel channels, bedrock fluting and escarpment noses, Kingston area, Ontario. Each fluting symbol represents many tens of individual fluting ridges

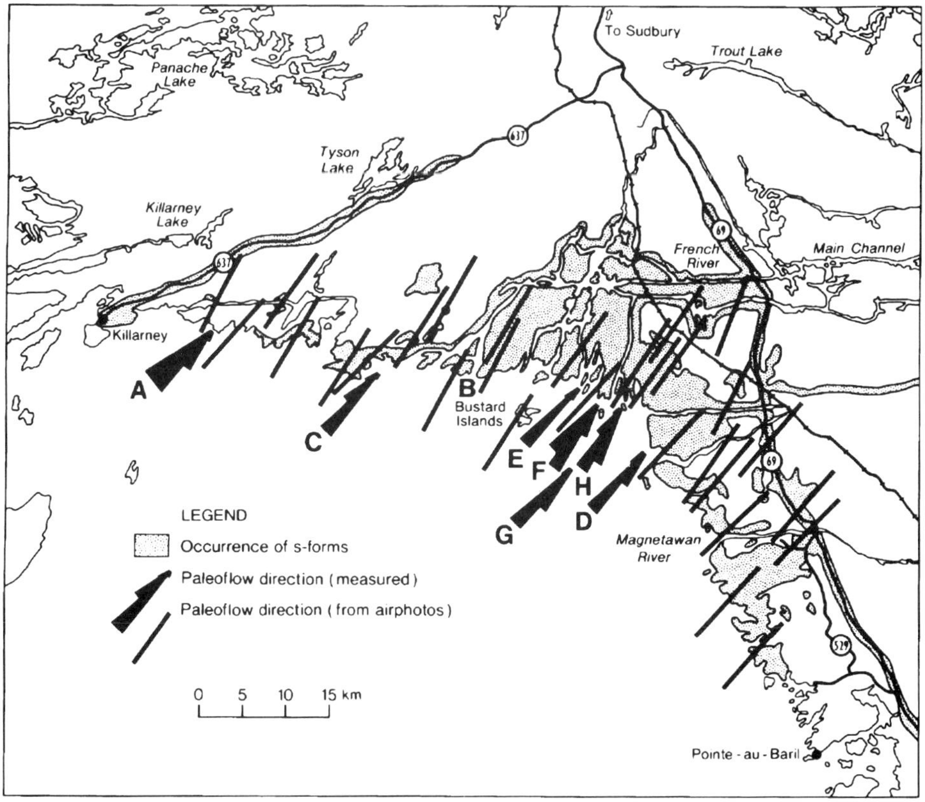

FIGURE 7.26 Distribution of s-forms, northeast section of Georgian Bay, Ontario, showing paleoflow directions obtained from direct measurement at field sites and from air photographs

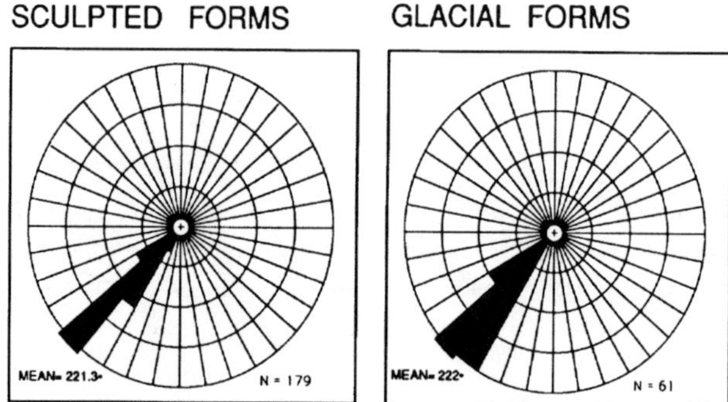

FIGURE 7.27 Circular histograms of flow directions estimated from s-forms and glacial forms (striations and crescentic fractures), mapped area of Figure 26, Georgian Bay, Ontario

Using these estimates for depth and velocity and 70 km as the minimum width derived from the map of the erosional marks, we obtain a conservative estimate for Q of 0.4 to 0.7 $\times 10^7 \text{ m}^3/\text{s}$. Draining at this rate, Lake Superior would empty in about 20 days!

DRUMLINS

A meltwater origin for drumlins is much easier to accept in light of the conclusions on regional-scale floods drawn from bedrock erosional marks. But the meltwater model did not evolve this way; conclusions on drumlins came first and were followed by the discovery of regional fields of bedrock erosional marks. This is a source of some satisfaction because our conclusions on drumlins, which were received with a great deal of scepticism and even violent opposition (Shaw 1988b), anticipated this discovery.

As mentioned in the introduction, features referred to collectively as drumlins may be genetically quite different. Some drumlins appear to be erosional and probably evolved around obstacles in the bed or at upstream-facing escarpments. These have classical

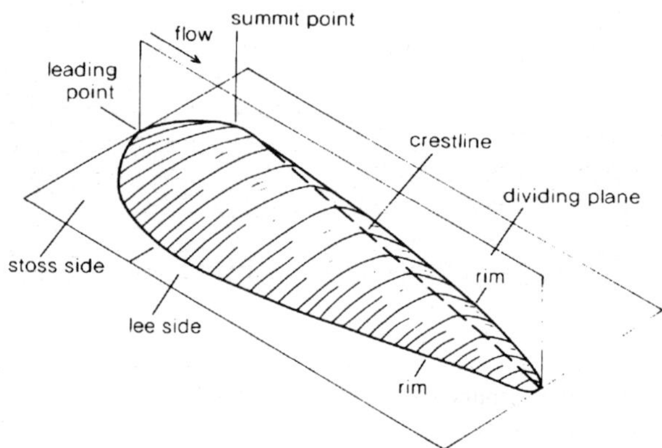

FIGURE 7.28 The classical drumlin with a steep, blunt stoss side and a tapered lee side

shapes with steep, blunt proximal ends and gently sloping, gradually tapering tails (Figure 7.28). They are probably part of a continuum with bluff, remnant ridges or rat tails defined by hairpin scours wrapped around the proximal end as one end member and isolated, streamlined hills as the other. I suggest the name *Beverleys* for these after the forms at Beverley Lake, Keewatin, NWT, illustrated by Prest (1983) (Figure 7.29). Fluting with parallel troughs or furrows is probably genetically related to Beverleys. For other drumlins, appearing like inverted erosional marks with spindle, parabolic and transverse asymmetrical shapes (Figure 7.30a and b), I propose the name *Livingstones* after the forms at Livingstone Lake, Saskatchewan, described by Shaw (1983). Thus,

FIGURE 7.29 Beverleys, Beverley Lake, NWT. Individual Beverleys are flat-topped ridges defined by pronounced hairpin scours that close in crescentic scours around the proximal ends of the ridges. Selected crescentic scours marked (A). Note that Beverleys occur at a variety of scales, a prominent smaller scale set appears about 2 km northwest of the north arrow. Flow to the northwest. This air photograph (T 300C-39) reproduced with the permission of Energy, Mines and Resources Canada

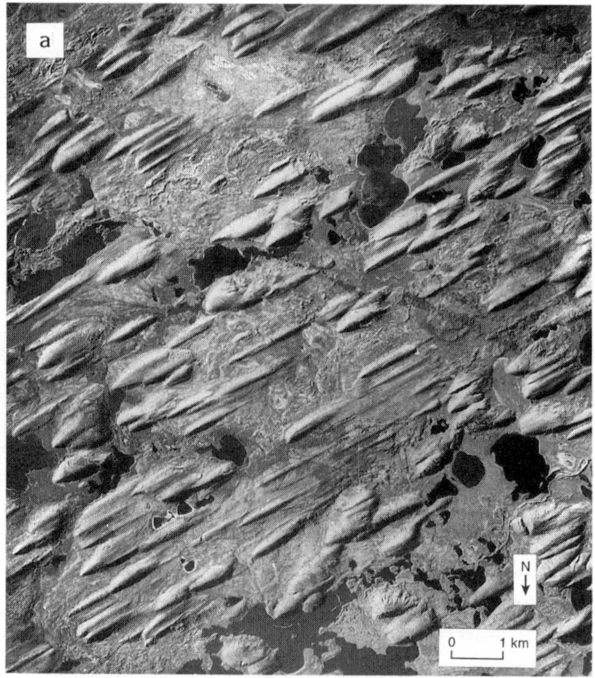

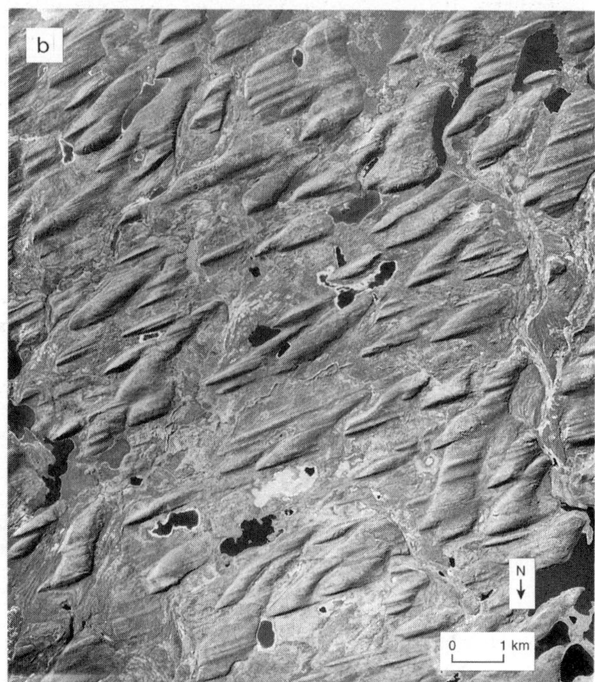

FIGURE 7.30 *For caption see opposite*

Beverleys and Livingstones may be combined with *Rogens*, named for forms at Lake Rogen, Sweden (Lundqvist 1969), in a set of subglacial bedforms.

Beverleys

Shaw and Sharpe (1987) noted that classical drumlins have very different shapes from those at Livingstone Lake; these classical drumlins resemble remnant ridges produced by water erosion and wind deflation and do not look like inverted erosional mark fills. They also noted that the cavity-fill explanation could not be applied generally to drumlins because neither bedrock, nor basal till, nor bedding truncated by drumlin slopes could be explained by accretion within a cavity. I remain surprised that the significance of these simple observations and conclusions escaped us for several years as we struggled to explain drumlins in terms of inverted erosional marks.

As soon as Beverleys or classical drumlins are seen as products of direct erosion, the problems of form, internal structure and composition disappear. Their form is erosional and each Beverley is simply composed of material that predates the erosional event. This could be bedrock, mineral sediment, or even organic deposits. Others have suggested an erosional origin for drumlins (cf. Gravenor 1957) and there is nothing new in these conclusions. But, whereas, in the past, as with fluting, it was assumed that ice was the erosional agent, now we must also entertain the possibility of meltwater erosion.

It is unlikely that there is a significant internal difference between Beverleys eroded by ice (cf. Gravenor 1957), by subglacial deformation (Boyce and Eyles 1991), or by meltwater (Shaw and Sharpe 1987). Consequently, choosing between these mechanisms, without actualistic observation, must be on the basis of form. For example, the characteristic form of scour around Beverleys plays an important role in our interpretation.

Some Beverleys are relatively flat-topped with accordant surfaces. They occupy most of the surface area of the field and are separated by relatively narrow furrows that wrap around their upstream ends (Figure 7.29) (Shaw and Sharpe 1987). The ridges are commonly *en echelon* and upstream furrows commonly run into, but do not cross, the crescentic part of furrows wrapped around downstream Beverleys (Figure 7.29).

Satellite images show Beverleys and their associated furrows, with lengths of tens of kilometres (Figure 7.31), that are identical in form to rat tails on a scale of millimetres or centimetres (Figure 7.32). Identical wind-eroded ridges with a similar range in scale, the larger ones being yardangs, are found on Earth and Mars (Figure 7.33) (Greeley and Iverson 1985). Both the wind-eroded features and Beverleys cluster downstream from upflow-facing escarpments with convex, sculpted noses (Shaw 1994).

FIGURE 7.30 (*opposite*) (a) Parabolic and spindle Livingstones, Livingstone Lake, Saskatchewan. Parabolic forms commonly show lateral ridges, counterparts of lateral furrows in s-forms. Unlike classical drumlins, spindle and parabolic Livingstones have sharply pointed stoss ends. Air photograph (A-14509-77). Flow from bottom left (northeast). (b) Transverse asymmetrical Livingstones, Livingstone Lake, Saskatchewan. The asymmetry in this part of the field is left-handed (viewed downflow, the bulk of each landform lies to the left of an axial plane drawn through the leading point). Spindle forms are superimposed on transverse asymmetrical Livingstones. Flow from bottom left (northeast). Air photograph (A-14509-5). Air photographs reproduced with permission of Energy, Mines and Resources Canada

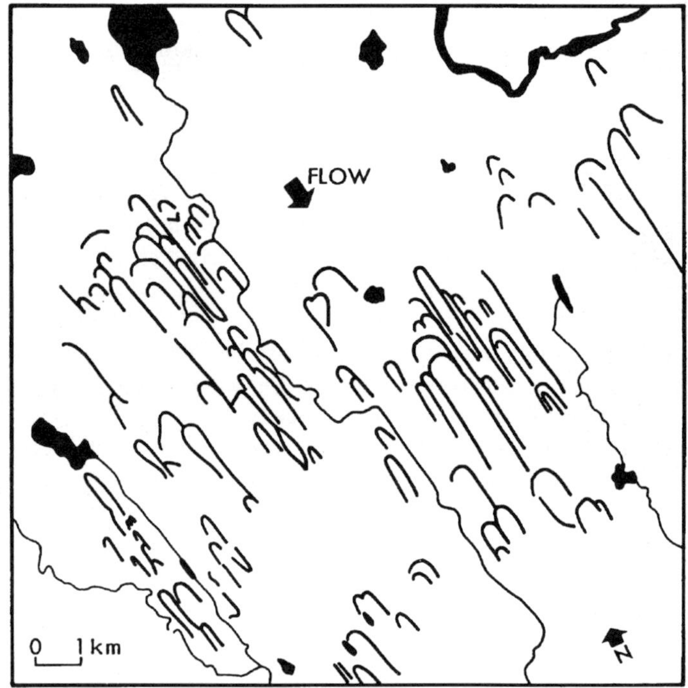

FIGURE 7.31 Hairpin scours sketched from Landsat 5 TM 19-25 Band 4. The centre of the area depicted is about 40 km northwest of Joutel, Quebec

FIGURE 7.32 Hairpin scours around chert clasts in Ordovician carbonate bedrock, Marysville, Ontario. Flow from right

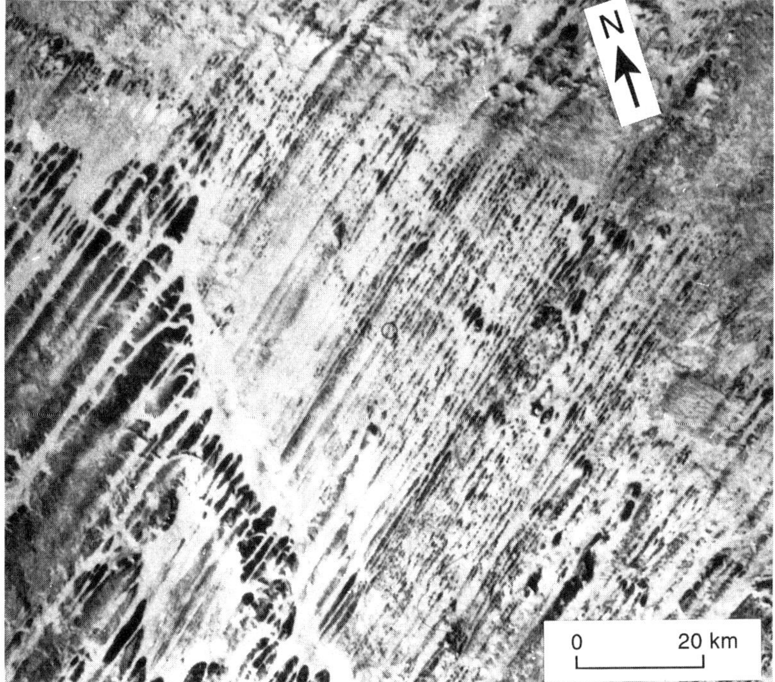

FIGURE 7.33 Landsat picture of yardangs (dark ridges) in Paleozoic rocks of the Borkou region, Chad. Formative wind from the northeast. Note the preferential development of yardangs at escarpments facing into the wind, prominent escarpment noses, and hairpin scours (light areas) wrapping around the stoss end of yardangs

The hairpin-shaped scours wrapped around Beverleys are best explained by the erosive action of horseshoe vortices in sediment-laden turbulent fluids (Figure 7.34) (Shaw and Sharpe 1987; Sharpe and Shaw 1989; Shaw 1994). These vortices form in flow boundary layers as a result of flow separation where fluid impinges on bluff obstacles (cf. Dargahi 1990). Like Beverleys and associated scours, horseshoe vortices are reported over an enormous range of scales. Details of scour morphology, the absence of striation where furrows are eroded into bedrock, and the extremely low Reynolds Numbers for ice flow (Shaw 1994) preclude formation of hairpin scours by ice. By contrast, wind and flowing water easily attain the form Reynolds Numbers ($> 10^4$, Allen 1982) necessary to generate erosive horseshoe vortices at the scales of observed hairpin furrows. As well, the morphology of Beverleys and yardangs corresponds in detail to that expected for erosion by these vortices (Shaw 1994).

Not all erosional drumlins have associated hairpin scours and many are relatively isolated, separated by broad scoured areas, commonly with exposed bedrock. These isolated forms are generally more streamlined, which probably explains the absence of hairpin scours: there is no flow separation in front of the leading face of streamlined forms. Both remnant hills in loess eroded by the Channeled Scabland floods and eolian erosional ridges, yardangs, show this streamlined form with lemniscate plan shape (Figures 7.35 and 7.36) (Baker 1978, fig. 7.15; Greeley and Iverson 1985, fig. 4-21).

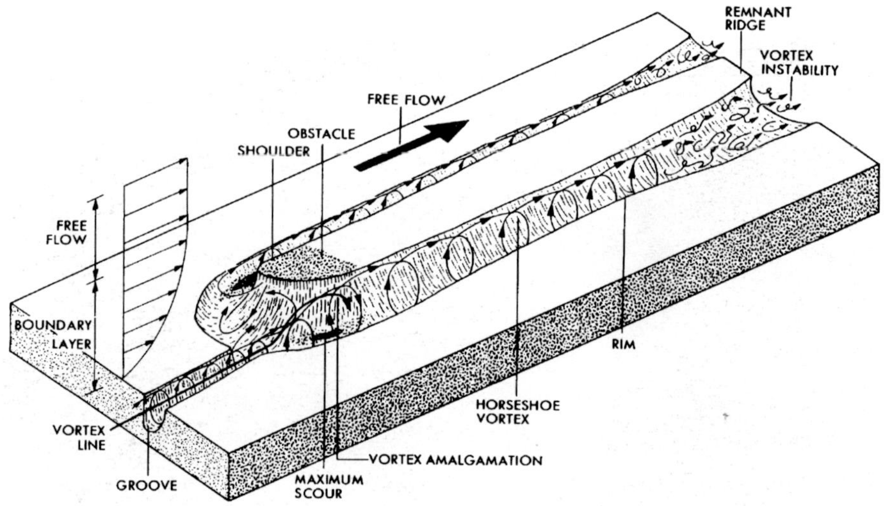

FIGURE 7.34 Erosion of a hairpin scour by a horseshoe vortex. Note how a vortex from an upstream hairpin amalgamates with the horseshoe vortex such that hairpin scours do not cross-cut one another

There seems to be a sequence in nature from hairpin scour around obstacles to isolated streamlined ridges that evolves as scour progresses (Figure 7.37). This suggests that landforms or bedforms in this continuum may be called Beverleys; the immature features are defined by hairpin scours and the mature ones appear as isolated, streamlined ridges without hairpin scours.

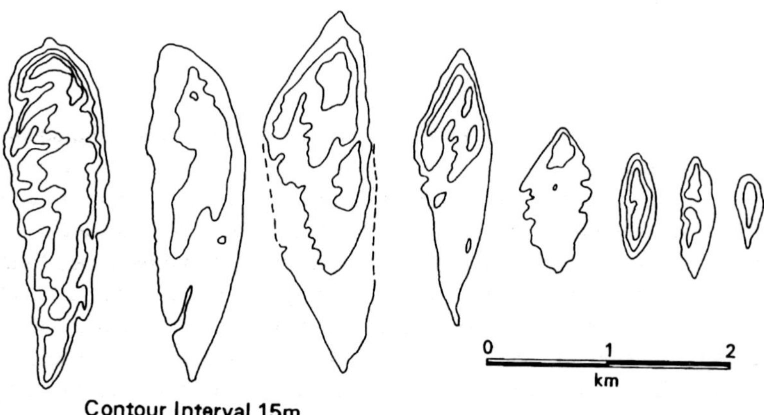

FIGURE 7.35 Contour maps of residual hills in loess, Channeled Scabland, Washington (after Baker 1978, fig. 5-10). Flow from top

FIGURE 7.36 Oblique aerial view of yardangs, Cerro Yesera, Peru. Prevailing wind towards viewer (after McCauley et al. 1977). Note the longitudinal flutes between the large yardangs

Livingstones

Landforms in a second set, widely recognized as drumlins and commonly used in textbook illustrations, have little resemblance to classical drumlins. These belong to the class of Livingstones. Spindle and parabolic Livingstones are pointed and narrower at the upstream end and flare distally. They are usually asymmetrical about a flow-wise long axis (Figure 7.30a). Proximal slopes of parabolics are steep and the base of the slope is sharply defined. Their gentler distal slopes merge gradually with the surrounding land surface. The crests of transverse asymmetrical Livingstones lie across the flow and the bulk of each landform lies to one side of a flow-wise axis drawn through the leading point (Figure 7.30b). Sets of these landforms show the same sense of asymmetry (left [right] if the bulk of the form is to the left [right] of the axis for a viewer looking downstream) over extensive areas. Their stoss slope is steep and commonly displays sharp offsets. Smaller-scale spindle Livingstones are common on their gentle distal slopes which merge imperceptibly with the surrounding ground. Whereas parabolic and transverse

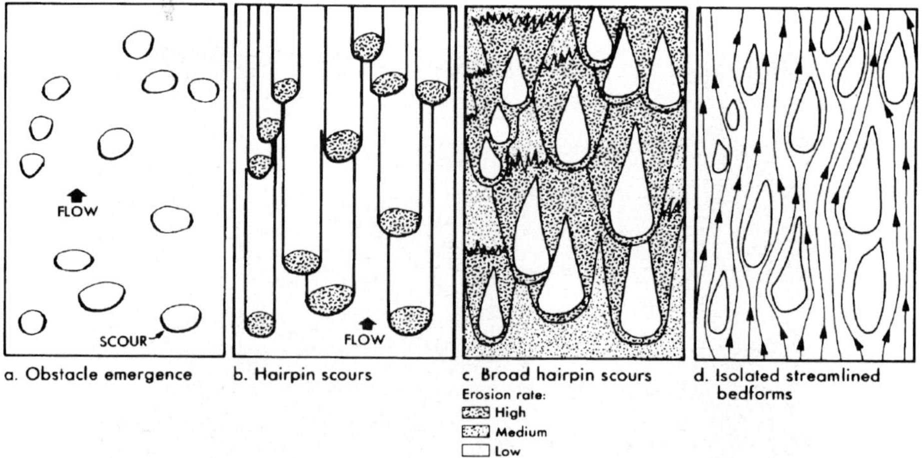

FIGURE 7.37 Sequence of erosional features formed after obstacles are created by differential erosion in a turbulent flow (a). Hairpin scours represent a period of horseshoe vortex formation around bluff obstacles (b). As the hairpin scours broaden the remnant ridges become more streamlined, tapering downflow (c). In the final stage (d), with full streamlining and no flow separation, hairpin scours are removed. At this stage, erosion is controlled by acceleration of flow over and between upstanding residual bedforms

asymmetrical Livingstones are found separately in extensive subregions of bedform fields, spindles are widely distributed over whole fields.

These forms inspired the original fantasy that drumlins result from meltwater sheet floods (Shaw 1983). The form analogies with other erosional marks have been refined over the years (Shaw and Kvill 1984; Shaw et al. 1989) and the classification and interpretation of subglacial s-forms (Kor et al. 1991) further supports a unified meltwater theory for positive and negative subglacial landforms (Figure 7.3).

Other observations allow us to reason beyond form analogies: sedimentology indicating deposition from hyperconcentrated flows, internal architecture conformable with external form, and dominance of local clasts in Livingstones are as expected for rapid deposition in cavities where flow velocities decelerated abruptly (Shaw et al. 1989). Other attempts to explain Livingstones in terms of wholesale subglacial deformation of ridges up to 80 m in height (Boulton 1987) account for neither the diversity of form nor the documented sedimentology of Livingstones in northern Saskatchewan (Shaw and Kvill 1984; Shaw et al. 1989).

Rogens may be related to Livingstones where their internal structure indicates deposition in cavities (Fisher and Shaw 1992). In this case, cavities in the ice were transverse forms like inverted ripples (Ashton and Kennedy 1972). Indeed, a photograph of inverted ripples breaking through the upper surface of river ice provides an almost perfect replica of the plan view of Rogen moraine (Figure 7.38).

Despite the confidence expressed in the meltwater model, some drumlins and perhaps some Rogen moraines do not fit neatly into its framework. These landforms show the shapes of Livingstones but contain a variety of sediment including basal till (Hanvey 1988). They are evidently erosional, though their form mimics Livingstones. Shaw

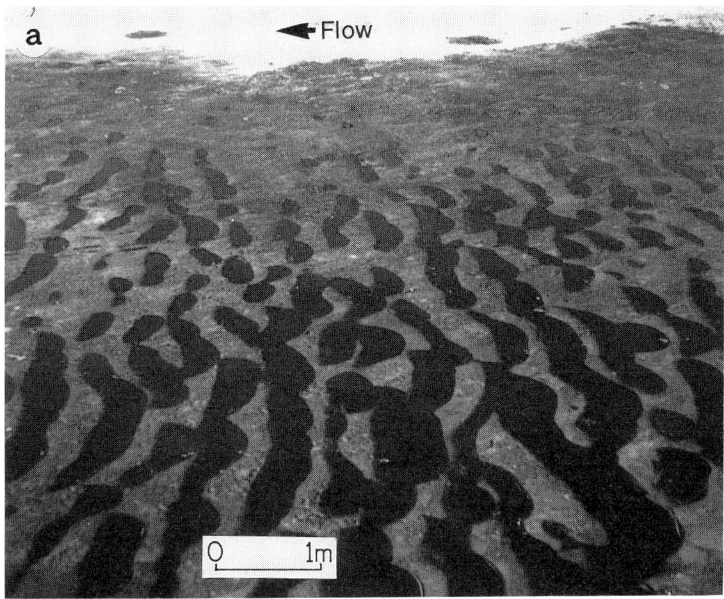

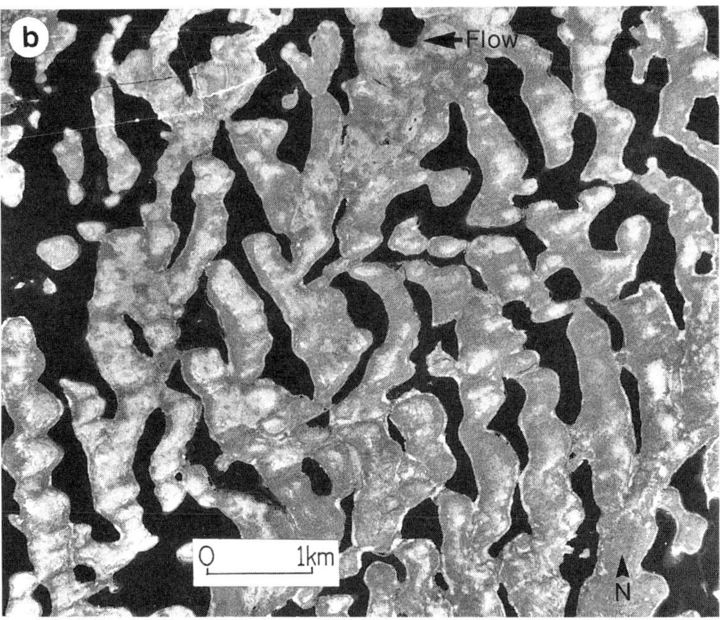

FIGURE 7.38 (a) Ripples cut upwards into the underside of river ice breach the surface in areas of exposed water (dark). Photograph courtesy of G. Ashton. (b) Rogen moraine, Boyd Lake area, NWT. This air photograph (A-14887-105) reproduced with permission of Energy, Mines and Resources Canada. The Rogens (light areas) are interpreted as cavity fills and are considered to be morphologically equivalent to the truncated ripples (dark areas) in (a)

(1994) suggests that they result from the effect of cavities in the overlying ice on the bed shear stresses of the meltwater sheet. Where bed erosion was general and sediment transport was below capacity, erosion would have been least below cavities. Residual ridges, shaped like the overlying cavities, would have formed there. It would be very difficult to differentiate cavity-fill and cavity-controlled erosional drumlins without details of their internal structure. It is, therefore, most convenient to treat both as Livingstones. This draws a morphological, rather than a genetic, distinction between Beverleys and Livingstones. A genetic implication is that Beverleys are controlled by obstacles to flow on the substrate and Livingstones are controlled by inverted erosional marks in the ice bed.

Discharge estimates

As for s-forms, discharge may be estimated for Beverleys and Livingstones assuming that their formation is by meltwater. The Livingstones of the Livingstone Lake field, northern Saskatchewan are up to 80 m high and boulders up to 3 m in diameter were transported by the flow event which formed the field. The Livingstones are an integral part of a much larger pattern of bedforms for which coherent flowlines can be drawn. As well, the field contains areas with Livingstones of distinct shape which merge into zones with different shaped drumlins; patterns are transitional rather than cross-cutting. This evidence is taken to indicate that Livingstones in a single field were produced by a flow at least as wide as the field. The same argument was presented earlier for s-forms around Georgian Bay. Assuming an extremely conservative mean velocity of 10 m/s, a conservative mean depth of 40 m, and a minimum flow width of 150 km gives an estimated instantaneous discharge of $6 \times 10^7\ m^3/s$.

Also by assuming that the volume of ice melted by the meltwater flow was three times the volume of Livingstones, energy considerations allow the total volume of meltwater producing the Livingstone Lake field to be estimated (Shaw et al. 1989). This turns out to be about $8{\cdot}4 \times 10^4\ km^3$, enough to raise global sea level by about 23 cm.

CONTINENTAL-SCALE IMPLICATIONS

The estimated discharges of meltwater events which are thought to have formed the Livingstones of northern Saskatchewan and the s-forms around Georgian Bay are both of the order $10^7\ m^3/s$ with a water volume of about $8.4 \times 10^4\ km^3$ required to form the Livingstones. These enormous discharges are almost certainly underestimated because the flow widths used in the calculations are very conservative. Floods of this magnitude are unlikely to have been generated locally and are expected to have had geomorphological and sedimentological effects over vast distances. This expectation provides a challenging test of the meltwater model: if a continuous sequence of landforms, inferred to be of meltwater origin, could be traced upstream and downstream of the Livingstone Lake drumlin field, it would mark the expected flow path. Flowlines derived from landform trends would be approximately synoptic. Thus, the hypothesis has testable predictive capabilities.

Landforms appearing in the meltwater model (Figure 7.3) and mapped from satellite images, air photographs and digital elevation models indicate just such a flow path from the Northwest Territories, through the Livingstone Lake area, northern Saskatchewan (Shaw and Kvill 1984), southwestwards into Alberta, through the Athabasca fluting field and southwards to the United States border (Figure 7.39) (Rains et al. 1993). Selected landscapes with prominent landforms along this flow path are illustrated in Figure 7.40. The flow path is merely a thin filament taken from a much broader flow field (Figure 7.39); the flow discharge and total volume must have been at least an order of magnitude higher than the estimates given here. This postulated meltwater outburst flood is the Livingstone Lake event of Rains et al. (1993).

A somewhat different approach may be taken regarding the evidence for the Algonquin outburst event in Quebec, Ontario and New York State. Shaw and Gilbert (1990) mapped the southern part of this event. Flowlines derived from drumlins mapped by Prest (1968) illustrate a fan-shaped flow extending from Labrador/northern Quebec southwards to southeast Ontario (Figure 7.39). Drumlins in northern New York State extend this flow path (Flint et al. 1959). By the meltwater interpretation of subglacial landscapes, Beverleys, Rogens, escarpment flow noses and reentrants, tunnel channels, bedrock fluting and other s-forms record the passage of this flow event. Flowlines of the Ontarian event cross-cut those of the Algonquin event (Shaw and Gilbert 1990).

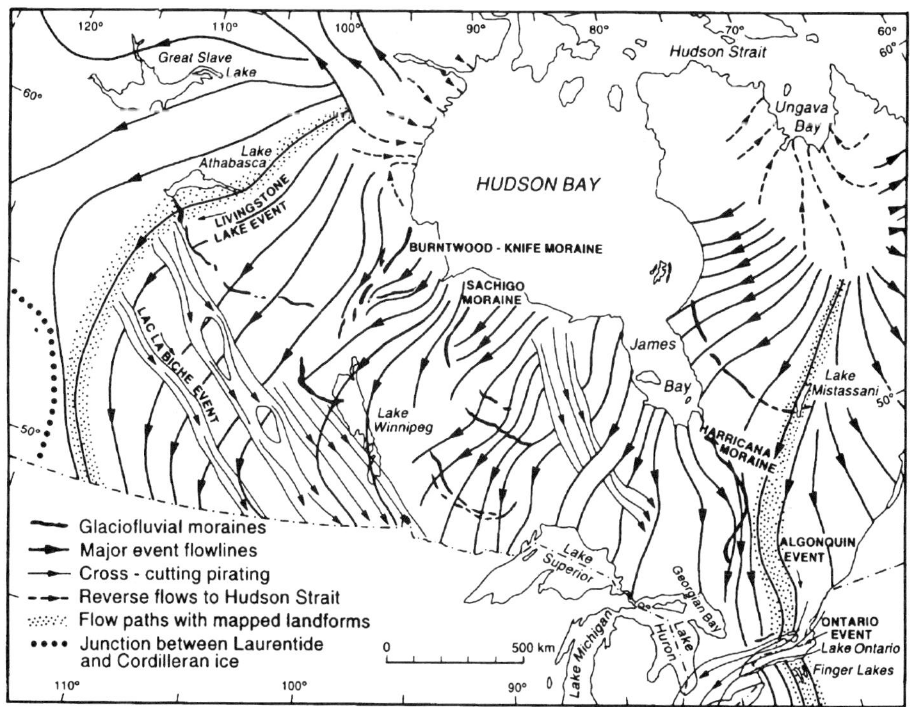

FIGURE 7.39 Generalized flowlines (water or ice?) for the Laurentide ice sheet derived from drumlins mapped by Prest (1968). Of particular note are the convergence of flowlines in the area around the Harricana moraine, the cross-cutting nature of the "Ontarian event", and the secondary northwest to southeast flow tracts (e.g. the Lac la Biche event) on the Western Plains

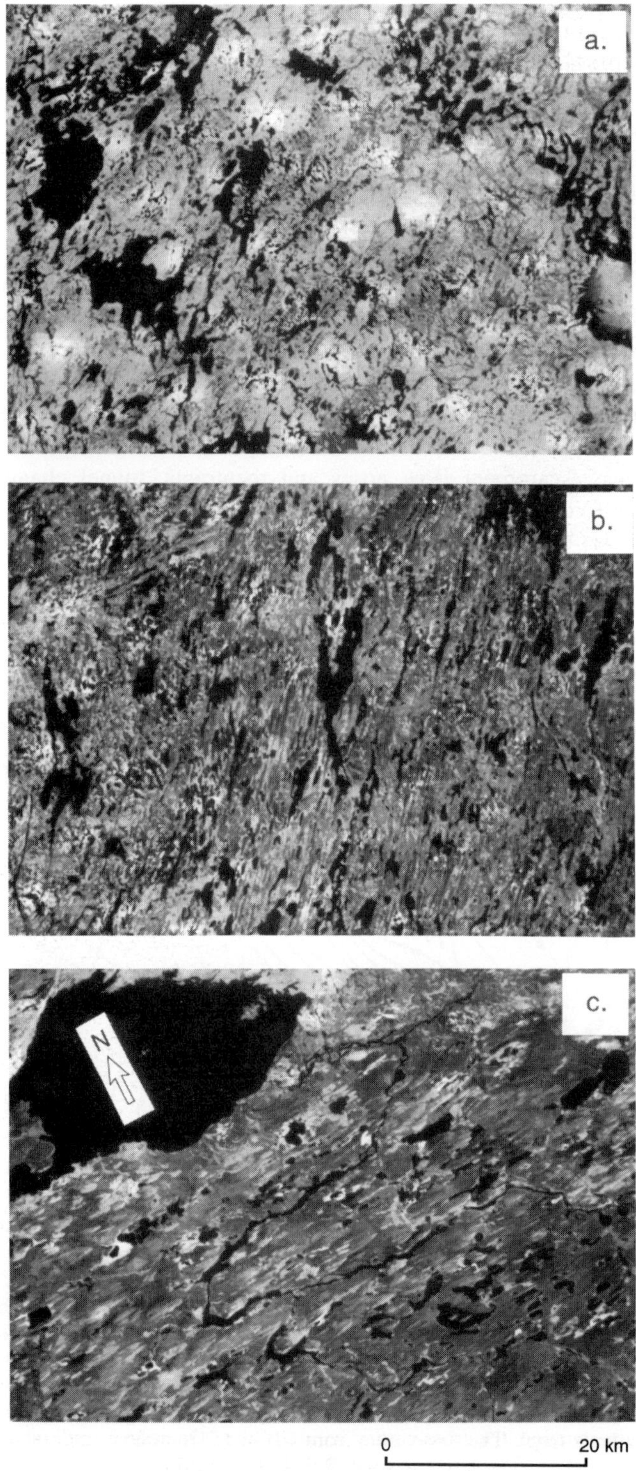

FIGURE 7.40 *For caption see opposite*

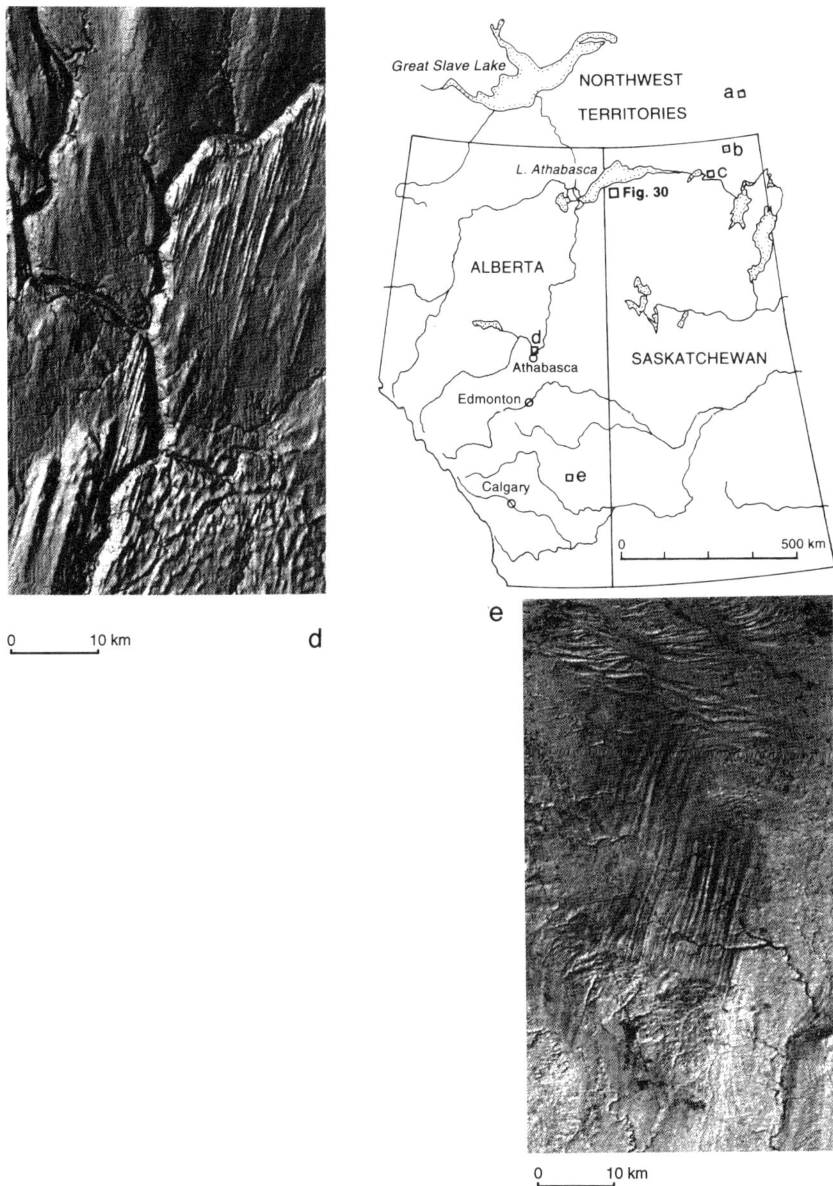

FIGURE 7.40 (*and opposite*) Landforms along the Livingstone Lake event flow path (Figure 7.39). Enhanced Landsat TM images: (a) mainly transverse bedforms (Rogens, see Figure 7.38b for a larger-scale illustration of these bedforms); (b) extremely long Beverleys (right), Rogens (left); (c) mainly parabolic Livingstones with superimposed fluting, located upflow of the Livingstone Lake drumlin field (Shaw and Kvill 1984). (d) digital elevation model showing large-scale fluting downflow from an escarpment (bottom left) and from upstream-facing tunnel channel walls (see Figures 7.20 and 7.21); and erosional marks resembling muschelbrüche and sichelwannen (bottom right). (e) Enhanced Landsat TM snow scene showing subtle fluting (centre) and cross-cutting channels with residuals (top). The flood tract from (d) to (e) is mainly scoured to bedrock with solonetzic soils (Rains et al. 1993)

Meltwater may also have eroded larger features. In the absence of the meltwater hypothesis, erosional troughs with depths measured in hundreds of metres, widths of the order of kilometres and lengths of tens of kilometres are attributed to glacial scour. For example, this interpretation was the only reasonable way to explain the overdeepened and enclosed rock basins of the Finger Lakes, northern New York State. Mullins et al. (1989) gave a fascinating account of the debate on their origin. The lakes occupy valleys cut through the Allegheny scarp and are actually part of a complex anabranched channel system with streamlined, interchannel residuals (Figure 7.41). The only end moraines in the distal parts of these valleys are the so-called Valley Heads moraines, composed of coarse sorted sediment. These moraines lie *within* the major troughs and, consequently, must have been deposited after the troughs were cut. Thus, the conventional explanation that the troughs were cut by preferentially eroding ice streams (cf. Mullins and Hinchey 1989) requires that these ice streams overcame a substantial reverse gradient which is expected to have caused compressive flow. They must also have followed an anabranching valley system with numerous reaches at high angles to flow and, in places, scoured below sea level. The ice streams must have terminated in a drainage system that was not ponded but carried meltwater to the oceans. Finally, moraines in till did not accumulate at the ice margin. Given that modern ice streams are flanked by sluggish ice not deeply entrenched in bedrock valleys, their flow is extending, not compressive, they terminate in ice shelves, and they are thought to transport huge volumes of debris by subglacial deformational processes, it is unlikely that ice streams carved the Finger Lakes.

By contrast, enormous discharges of subglacial meltwater with a substantial hydrostatic head and a high potential gradient across the escarpment could have eroded the anabranched valleys and transported the resultant sediment through proglacial valleys to the ocean. Indeed, deep scouring with potholes and residual islands in bedrock testify to extraordinary erosion in the Susquehanna Valley, Pennsylvania, which, in the outburst flood interpretation, would have carried flood waters of the Algonquin event.

All of the features of the Finger Lakes attributed to ice action, hanging tributary valleys, streamlined hills, deeply scoured basins, and relatively undissected valley-sides are just as well explained by catastrophic meltwater flow (Gilbert and Shaw 1994). Moreover, there are features that are difficult to explain if the valleys were eroded principally by ice streams: the anabranched valley system with reverse gradient, the absence of till moraines, and the stratified sediment in the Valley Heads moraines.

A second lake system, Lakes Mistassini and Albanel, Quebec, also lies on the Algonquin flow path. It is preferentially eroded into Proterozoic sedimentary rocks which are less resistant than surrounding crystalline basement rocks (DiLabio 1981). Elongated, discontinuous troughs, some in excess of 500 m deep, wind along the bed of Lake Mistassini (Figure 7.42). Similar, but less well-developed troughs dissect the floor of Lake Albanel. A slope curvature map accentuates the deep channels in the lakes and illustrates numerous terrestrial channels that begin and end abruptly (Figure 7.43). These channels can only be tunnel channels submerged beneath a sheet flow or connecting broader zones of subglacial meltwater flow. The discontinuous, elongate troughs in the lakes must have been eroded beneath a flow that was much wider than the troughs themselves. The simplest conclusion is that the troughs in the lakes and the terrestrial channels were both submerged in a broad sheet flow.

FIGURE 7.41 Digital relief model of the Finger Lakes area, New York State, based on 3-second arc digital elevation data. Note the deep reentrant valleys of the Finger Lakes breaching the escarpment and dividing downflow into an anabranched valley system. This map depicts only the topography and does not indicate the full depth of the bedrock troughs

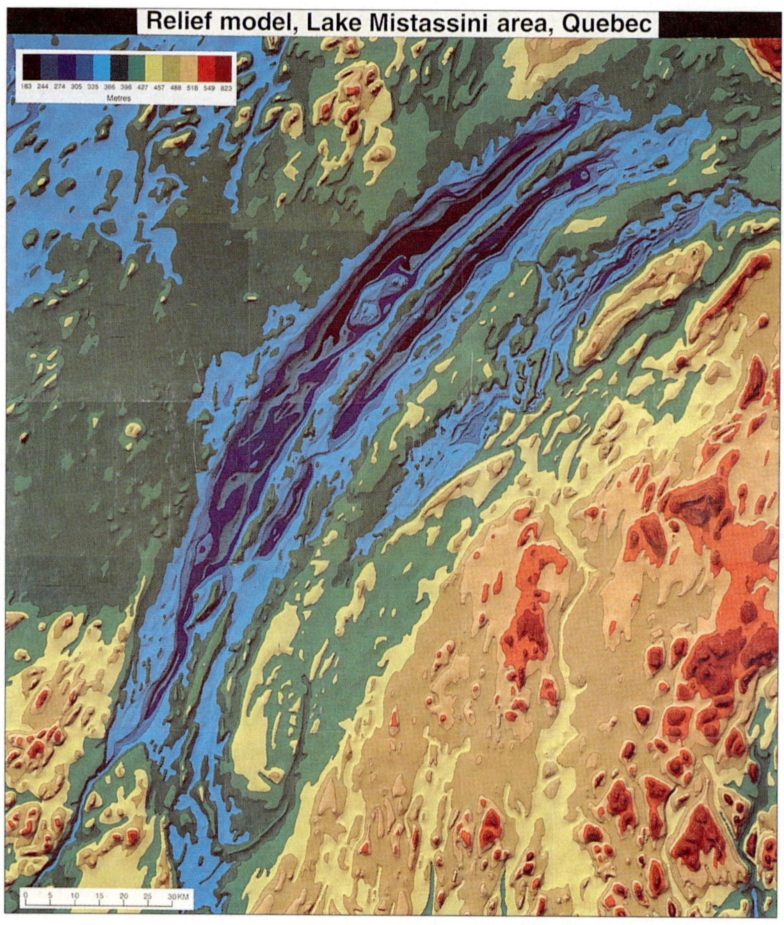

FIGURE 7.42 Digital relief model of the Lake Mistassini area, Quebec. Digitised from NTS 1:50 000 scale maps and bathymetric surveys provided by Département des Resources Hydrauliques, Quebec. This map shows the full depth of the lake basins and illustrates deep erosional troughs in their beds

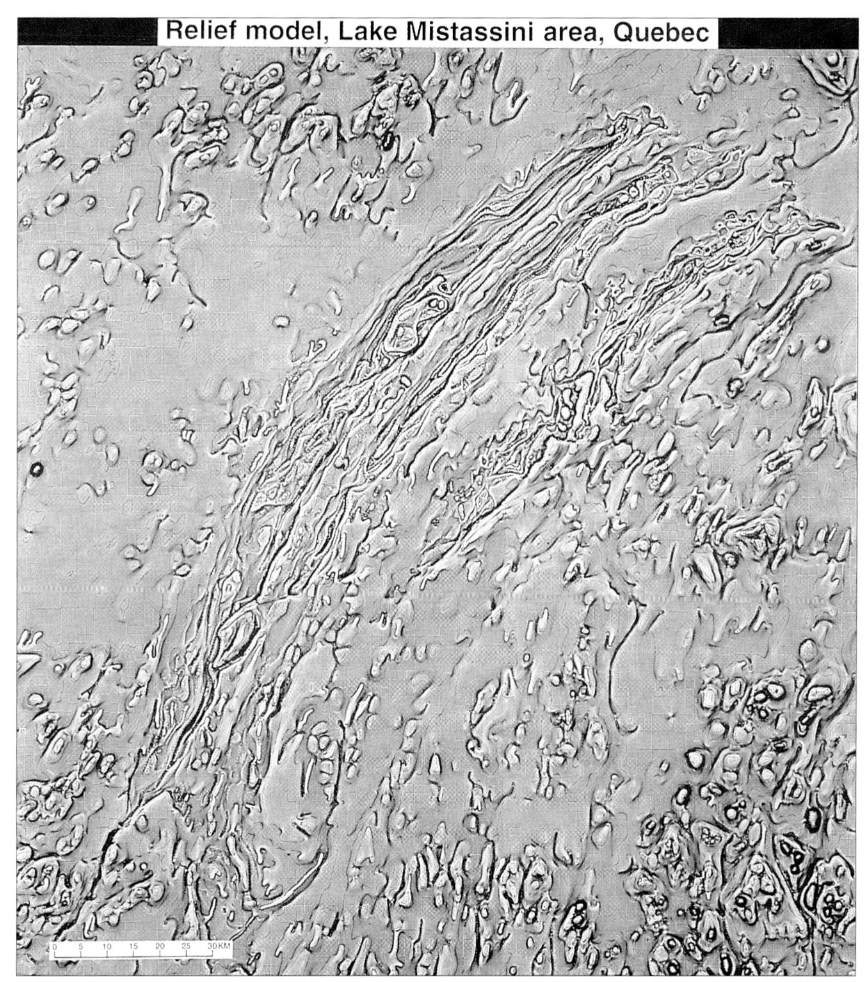

Figure 7.43 Slope curvature map of the Lake Mistassini area, Quebec, based on the digital elevation data of Figure 7.41, using a continuous grey scale. Areas of surface convexity appear light and areas of concavity are dark. This technique accentuates the troughs in the deep depressions and the numerous, short channels (tunnel channels) which start and stop abruptly

Significantly, the spectacular potholes on Colline Blanche (Gagnon 1988), located less than 10 km from the eastern shoreline of Lake Albanel (Figure 7.14), are interpreted to have formed in a broad flow that completely submerged the hill. As well, the lakes lie within an extensive field of Beverleys and there are Rogens in the vicinity (Bouchard 1989). Thus, in the meltwater model, the lakes represent preferential scour beneath a water sheet that formed the potholes on Colline Blanche and the subglacial bedforms. Meltwater in this sheet flowed to the ice margin, which lay beyond the Finger Lakes, eventually reaching the ocean. Given the amount of meltwater and the magnitude of the erosion in the Finger Lakes and Lakes Mistassini and Albanel, it seems reasonable to speculate on the role of meltwater erosion in the formation of the Great Lakes Basins.

THE MELTWATER SOURCE

The meltwater hypothesis has been discussed so far largely in terms of evidence by which it is validated. A hypothesis that began as fantasy is made real, or as real as a hypothesis of past events can ever be, through the correspondence of observations and expectations. However, detractors almost always raise the question of the meltwater source, suggesting that our failure to give a specific location of the meltwater reservoir and concrete evidence for its existence presents insurmountable objections to the whole hypothesis. Bretz faced similar criticism when he first proposed a flood origin for the Washington Scablands (Baker 1978). Like Bretz, I argue that the evidence presented is evidence enough that a reservoir must have existed. As well, the middle-latitude Laurentide ice sheet would have had several potential storage sites and a more than adequate water supply.

Nevertheless, in the absence of direct evidence for the reservoir, we must go back to fantasy, to images of what might have been. This exercise can at best outline plausible reservoir locations and release mechanisms based on the deduced flow characteristics. In time, it may be possible to validate what, for now, is speculation. Even so, the speculation is constrained and guided by the meltwater model in the following ways. If the meltwater hypothesis outlines what actually happened, large volumes of meltwater must have been released catastrophically in outburst floods from beneath the Laurentide ice sheet. The largest flows were released when this ice sheet was still at or close to its maximum extent. Thus, release took place relatively soon after this maximum was reached.

Beverleys and flutings, which are eroded into glaciotectonized bedrock and thick sequences of glacigenic sediment, including lodgement and melt-out tills and subglacial glaciofluvial deposits (Figure 7.21) (Hicock 1988; Shaw 1994), are preserved in almost pristine subglacial landscapes. Hence, glacial tectonism and much glacial erosion, transport and deposition preceded the dramatic events that formed the landforms depicted in Figure 7.3. Afterwards very little geomorphological alteration took place over whole flood tracts. This last conclusion depends on contemporaneous formation of the Beverleys and flutings along a flow path. This is a point of contention and requires justification here.

This justification stems from systematically organized, regional-scale, subglacial landscapes. Fan-shaped drumlin fields, broadening towards the ice margin, are a common element of these landscapes in Canada and Finland. They are associated with eskers, "interlobate" moraines and arcuate end moraines. The whole assemblage may be idealized (Figure 7.44). There are a number of critical aspects to this organization. Drumlins extend beyond the arcuate moraines but do not cross the interlobate moraine. Although drumlins are arranged in large, fan-shaped fields, drumlins in one large fan do not cross-cut those of adjacent fans. Rather, neighbouring fields meet in a broad zone in which drumlins cannot be clearly identified. An "interlobate" moraine, for example the Harricana moraine (Veillette 1986; Brennand 1993), extends along the full length of this zone. Such moraines are invariably of glaciofluvial origin and are commonly discontinuous, being broken by gaps (Brennand 1993). Striations and feeding eskers converge on the "interlobate" moraines (Veillette 1986; Brennand 1993). The arcuate end moraines are also characterized by feeding eskers, although the eskers may stop short of narrow segments of the moraines (Fyfe 1990). Major changes in flow direction are indicated where bedforms of small fans cross-cut larger drumlin fans (Figure 7.39) (Punkari 1982).

Evidently, if one set of drumlins of neighbouring large fan-shaped fields does not cross-cut the other, both sets must have formed at the same time. If meltwater created the drumlins, then flow convergence in the zone between the fields must have produced extremely high flow discharges there. Such a convergence zone south of James Bay shows evidence of intense meltwater erosion: an anabranching channel system cutting

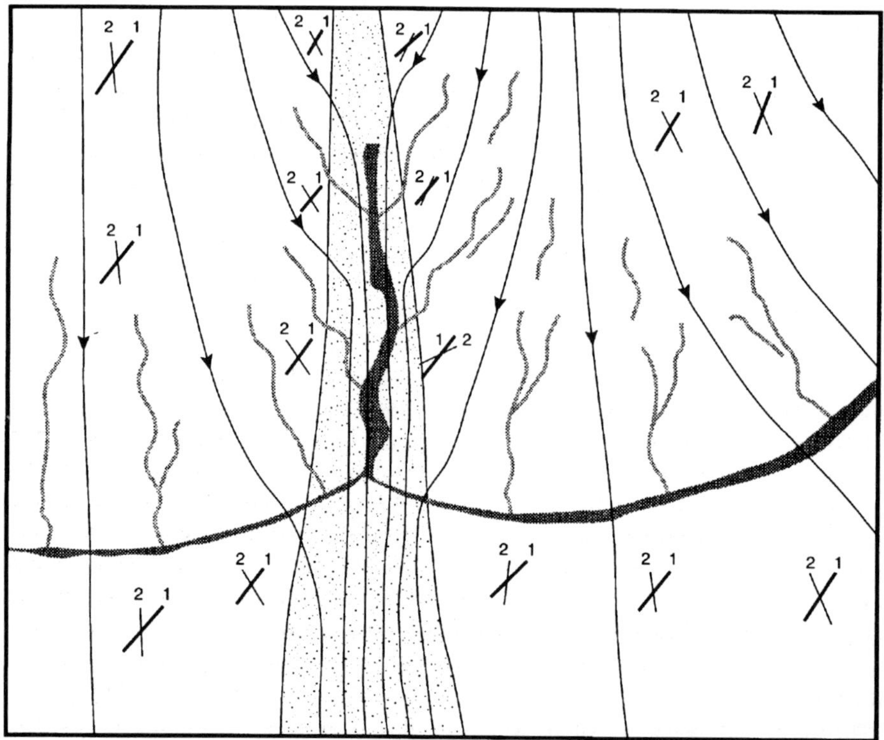

FIGURE 7.44 Idealized relationships between flowlines over drumlin fans and associated interlobate and arcuate glaciofluvial moraines

through the Abitibi Uplands passing southwards into the s-forms of Georgian Bay (Kor et al. 1991), large-scale s-forms around Kirkland Lake and Larder Lake, Ontario (Christian 1988), broad fields of fluvially eroded bedrock at Noranda-Rouyn, Quebec, and potholes perched high above bedrock valleys at Parc D'Aiguebelle (Charbonneau, personal communication, 1994). Just as the bedrock was scoured extensively, extreme thermal erosion of the ice would have ablated the roof and walls of the broad conduit (Röthlisberger 1972). Consequently, ice flow towards the conduit from the sides and from above would have produced a broad and deep supraglacial trough directly over the conduit.

Supraglacial drainage towards this trough and a connection to the ice bed through moulins would have ensured continuing flow through the conduit after the sheet-flow stage. Continued ablation of the conduit walls and roof would have maintained the surface trough and caused ice and subglacial meltwater to flow with a component towards the conduit. Thus, striations, which cross-cut those of earlier, regional ice flow, converge on the trough, as did the tunnels in which large eskers formed (Veillette 1986; Brennand 1993). Finally, deposition in the large conduit and the esker conduits produced the "interlobate" glaciofluvial moraine and eskers. This interpretation corresponds closely to the views of Veillette (1986) and Brennand (1993) that the Harricana "interlobate" moraine is a large esker. Brennand (1993) also concluded that it was deposited, as a whole, synchronously in a continuous conduit. Gaps in the deposit probably relate to erosion by superimposed supraglacial streams (Brennand 1993).

Thus, the ice lobes were a consequence of the sheet flow geometry. Drainage and sediment transport towards the ice front, where it terminated in large proglacial lakes and remained in a stable position, resulted in large arcuate glaciofluvial moraines. Examples include the Salpausselkä, Cree Lake, Middle Swedish, Sakami, and Hartman moraines.

This reconstruction of events gives hints on the reservoir location and release mechanism. For example, it is difficult to imagine a release mechanism triggered at the ice margin that would have caused simultaneous outburst floods from beneath much of the ice sheet. By contrast, if the triggering mechanism came from near the ice-sheet centre, meltwater flow outwards in all directions is a possibility. Again, imagination, rather than observation, is needed to visualize how this might have happened.

In the following thought experiment, imagine that, after surging of ice streams and lobes, perhaps related to subglacial deformation processes (cf. Boulton et al. 1985), the ice sheet has a very flat profile with large supraglacial lakes (Figure 7.45). Similar, though relatively small, supraglacial lakes are reported on parts of modern ice sheets with low surface gradients (Mellor and McKinnon 1960; Echelmeyer and Harrison 1990; Russell 1993). If these lakes expanded and deepened by thermokarst processes to cover a large area of the Laurentide ice sheet, extending from near the ice margin to the Hudson Bay area, they would have had dramatic effects on the thermal regime. Surface ice temperatures would have been at 0°C and vertical temperature profiles would have been close to isothermal. If the Laurentide ice sheet was virtually stagnant over the Western Plains of Canada and had these thermal characteristics, melting at the bed by geothermal heat may have produced extensive melt-out till sequences with intratill glaciofluvial beds recorded on the Prairies (cf. Shaw 1982, 1987). Ponding of water would also have been likely in the interior parts of the ice sheet (Shoemaker 1992). Eventually, when the supraglacial lake connected to the bed, the enormous head and volume of meltwater would have caused a powerful subglacial flood to surge outwards towards the ice margin (Figure 7.45). It is important that the supraglacial lake and the subglacial meltwater system remained disconnected as the lake grew. If this had not been the case, catastrophic drainage of the required magnitude could not have happened. I suggest that drumlins, flutings, Rogen moraine and s-forms may have been created by such catastrophic outburst floods. This theoretical model provides explanations for some puzzling changes of direction in drumlin-forming flows. For example, the change in flow direction between the so-called Algonquin and Ontarian events (Figure 7.39) (Shaw and Gilbert 1990) in Ontario and northern New York State is readily explained in terms of changes in flow conditions for a single event. Imagine that, at first, the head of water in the supraglacial

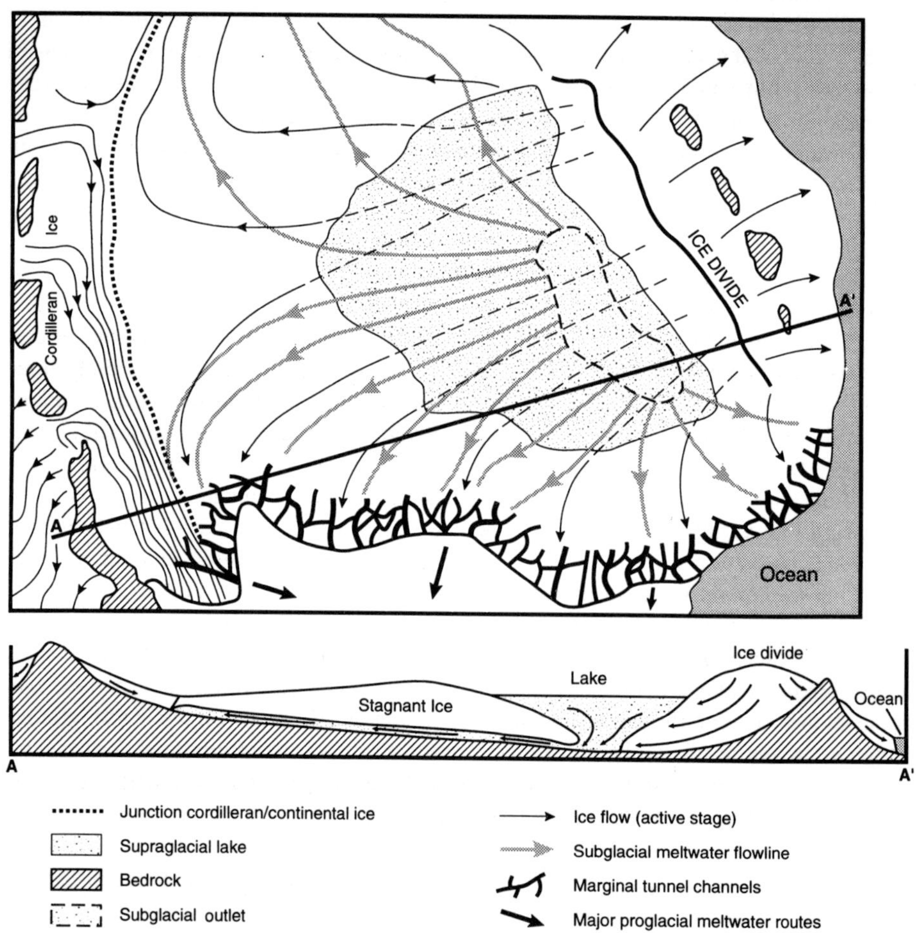

FIGURE 7.45 A hypothetical ice sheet, resembling the Laurentide, with a large area of stagnant ice and an extensive supraglacial lake. The cross-section captures a synoptic view following the connection of a large supraglacial lake with the subglacial hydrologic system. Estimated flowlines for the outburst flood resulting from this connection are shown. These flowlines would be diverted with time, producing cross-cutting bedforms, as the head of water in the lake became incapable of driving floodwaters through the higher outlets

lake was sufficient to drive escaping meltwater over the Allegheny escarpment into which it carved the Finger Lakes (Figure 7.41). The drumlin fields of the Algonquin event were formed at this stage (Shaw and Gilbert 1990). As the supraglacial lake drained, the head would eventually have fallen below the level required to drive meltwater across the escarpment. At that point, meltwater from the north was diverted south-southeastwards along the axes of Lakes Ontario and Erie — the Ontarian event of Shaw and Gilbert (1990). Thus, in this reconstruction, what were previously thought to be discrete events, were different stages of the same event.

A similar approach successfully explains cross-cutting subglacial landforms on the Western Plains of Canada. An initial southwesterly and southerly flow of meltwater over the plains continued as long as the head was sufficient to drive meltwater up-gradient along the Livingstone Lake route (Figure 7.39) (Rains et al. 1993). This was no longer possible when the head fell below the western outlet and the flow was rerouted southeastwards through the Lac la Biche fluting field to the Souris drainage in North Dakota (Rains et al. 1993). Similar piracy of the southwesterly flow over the Prairies produced numerous swathes of southeasterly trending fluting cross-cutting the initial southwesterly fluting (Figure 7.39). Thus, in this interpretation, cross-cutting relationships are more a product of falling meltwater head relative to outlet levels than of changing ice divides as suggested by Boulton and Clark (1990) and Clark (1993).

Why outflow was concentrated in fans remains to be explained. Presumably, subglacial drainage was controlled by the variable geometry of the gap between the substrate surface and the ice bed; meltwater would have been funnelled through zones of relatively high gap width and these would have concentrated flow from upstream, behaving like outlets from a lake. The lake analogy is not quite appropriate because, under high subglacial water pressure, several outlets may operate at different levels. Assuming that flow was funnelled through a small number of such "outlets" around a core area where the supraglacial lake system connected with the subglacial system, it is conceivable that it fanned out from these towards the ice margin (Figure 7.45). Shoemaker (1992) suggested a similar configuration for the subglacial parts of outlets from subglacial reservoirs.

A common criticism of the meltwater hypothesis for subglacial bedforms is that it invokes broad sheet flows that are theoretically unstable, being channelized quickly (Walder 1982). While the theory applies to steady-state, equilibrium conditions, the conditions invoked for the meltwater hypothesis are highly transient. Sheet flows are thought to have been relatively short-lived, though there was sufficient time for them to create an impressive erosional and depositional landscape, and eventually to have channelized as theory suggests (Shaw 1983; Brennand and Shaw 1995).

IMPLICATIONS FOR THE EVOLUTION OF THE LAURENTIDE ICE SHEET

The ways in which landforms are used as indicators of ice flow and in ice-sheet modelling are not consistent with the meltwater hypothesis. The conventional approach could be applied successfully to the preflood landscape in which striations, glaciotectonic structures, dispersion trains, till fabrics and glacial landforms record ice-sheet behaviour. If the interpretations given here are correct, the modern Canadian landscape, largely preserved from beneath the Laurentide ice sheet, reflects mainly the effects of meltwater flows. Perhaps the most dramatic change of interpretation involves the major ice divides (Dyke and Prest 1987; Boulton and Clark 1990).

For example, in the conventional view, the Keewatin area was covered by an ice dome with bedforms around this former ice divide indicating synoptic, diverging flow (Aylsworth and Shilts 1989). In the meltwater interpretation, the landforms indicating flow with a westwards component were formed first, by outburst floods from the Hudson Bay area. Those indicating an eastwards component of flow were formed later by meltwater escaping through Hudson Strait. Indeed, Aylsworth and Shilts (1989, map sheet 2) show cross-cutting landforms, in the area covered by the southeastern part of the map, suggesting late-stage eastwards flow.

This reconstruction holds promise as an explanation of dramatic and problematic marine deposits, the Heinrich layers — distinct layers in North Atlantic marine sediments rich in ice-rafted debris (Heinrich 1988). The layers are thought to mark a massive purge of the Laurentide ice sheet in which enormous discharges of icebergs were carried through Hudson Strait to the North Atlantic Ocean. Note that the iceberg discharge was but part of a more general cooling and freshening of surface waters in the North Atlantic. Bond et al. (1992) make it clear that the cooling and freshening accompanied by a reduction in the abundance of planktonic foramanifera were a composite of two processes, one being the iceberg discharge. The other process might well have been massive outburst flooding, preceding the collapse of the Laurentide ice sheet. Outbursts would certainly have caused cooling and freshening; turbidity underflows would have diluted cold bottom waters, causing them to rise to the surface. The timing of a prominent and extensive Heinrich layer at about 14 000 years ago (Bond et al. 1992) also accords with the estimated age of the major Laurentide outburst floods.

Expanding on floods as a mechanism for Heinrich layers, in the first outburst stage, flow was to the south, west and north from Keewatin and southwards from Ungava Bay (Figure 7.39). At this time, Hudson Strait must have been sealed by thick ice extending from Baffin Island to the Torngat Mountains of Labrador. Otherwise, the bulk of the meltwater would have escaped through the strait as a result of the flow efficiency provided by a shorter flow path and a lower outlet level to the North Atlantic. With drainage of the supraglacial reservoir, ice over Hudson Strait would, at first, have thinned as it was drawn down towards Hudson Bay. At the same time, sea-level rise caused by outbursts, as much as 10 m, would have destabilized the ice sheet on the continental shelf to the east (Blanchon and Shaw 1995). The Baffin–Torngat ice would have been drawn down further and the Hudson Strait seal would eventually break, giving rise to enormous outpourings of ice and water to the North Atlantic, the Heinrich events. At this stage, flows from Keewatin and through Ungava Bay reversed towards Hudson Strait.

These extensions of the meltwater hypothesis to ice-sheet modelling are preliminary. Nevertheless, they are consistent with the hypothesis and with reinterpretations of landform assemblages and marine sediments. The reinterpretation of the Heinrich layers, for example, better explains the short periods between events at the end of the late Wisconsinan than do those based on cyclic changes in glacier-bed thermal conditions requiring cycles of 7000 years (MacAyeal 1993). The timing of a major outburst before about 14000 years BP is consistent with the geomorphological evidence for an event sometime after the Laurentide ice-sheet maximum.

EXTRAGLACIAL EFFECTS

If there were huge meltwater outbursts, evidence for them must be recorded in the extraglacial river valleys that carried the floodwaters to the oceans. As well, there must be spectacular sedimentary records in the ocean basins. The floods must also have had dramatic effects on sea level, ocean salinity, isotopic composition, stratification and currents. In turn, coupling of the oceans and atmosphere implies that the floods may well have significantly influenced lateglacial climate. These considerations, which are wide ranging and worthy of a full-length paper, are commented upon briefly here.

The proposed meltwater floods must have followed the Mississippi, Susquehanna, and Hudson valleys. Each of these valleys has been deeply scoured (Mathews 1917; Fisk 1944; Thompson 1985). The river beds were eroded below the lowest glacial sea levels. This erosion below base-level cannot be explained by normal fluvial processes. By contrast, it is reasonable to assume that, where broad subglacial sheet flow was funnelled into a constricted valley, it became more efficient and eroded the valley floor and walls. Thus, the observed, deep scour is as expected in valleys thought to have carried floods.

At least 10^4 km^3 of sediment are estimated to have been eroded from Alberta alone during the Livingstone Lake event. This sediment may have been deposited as backwater deposits in tributary valleys of the Missouri and Mississippi rivers, to be subsequently reworked by the tributaries or by wind. Thick loess accumulations in the American Midwest may partly owe their origin to this sediment.

Even so, vast quantities of sediment and floodwater must have reached the Gulf of Mexico at the mouth of the Mississippi. What would have happened to this inflow? It is most likely that heavily sediment charged floodwater would have formed a direct density underflow. With relatively low sea levels, it would have plunged over the continental slope, accelerated and, perhaps, eroded the sea bed; erosion may even have extended beyond the slope. Deposition would have dominated, as the turbidity current density was reduced, by mixing with ambient sea water and by sediment deposition. An immense deposit is expected in deep water, offshore from the mouth of the Mississippi Valley.

Weimer (1989), using seismic sounding, recorded accretion of up to 4.5 km of sediment below present-day water depths of about 2000 m. There is over 10^6 km^3 of sediment in this submarine fan. Weimer (1989) recognized 17 sequences, several of which overlie erosional surfaces. They are capped by fine sediment and some contain thick beds with a chaotic seismic signal. These beds are interpreted to be products of gravitational mass movement. The coarse-grained components of the sequences are said to be submarine channel, levee and overbank deposits originating during low sea level, when sediment was transported to the shelf edge. The capping fine sediments are so-called condensed sections representing long periods of high sea level (Weimer 1989).

There is no reliable dating of this sequence: Weimer (1989) suggests that it goes back to the Pliocene, presumably on the assumption of steady-state sedimentation rates. Walker (1992) attributes the upper part of the sequence to the late Wisconsinan, but gives confusing dates for this. The fact seems to be that the deposit is not well-dated. Given the enormous discharges and sediment delivery expected for outburst floods, it is worth considering that much of the submarine accretion described by Weimer (1989) resulted from catastrophic meltwater release. The sedimentary characteristics are well-explained by periodic floods with cataclysmic erosion and deposition.

It is probably no coincidence that an enormous turbidity deposit, the Black Shell turbidite, dated at approximately 16 000 years ago, is found on the Hatteras Abyssal Plain off the mouth of the Susquehanna Valley (Pilkey 1988). Extensive tracts of large-scale fluting are also recorded beneath about 3000 m of water on the floor of the Labrador Sea off Hudson Strait (Hesse, personal communication). Both the Susquehanna Valley and Hudson Strait lay along proposed flood routes.

The turbidity underflows must eventually have become bouyant as they deposited sediment. At some point, they would have separated from the bed and risen to the surface in plumes. In this way, the water column would have become thoroughly mixed and the surface waters would have become cold, of relatively low salinity, and of relatively light oxygen isotopic composition, strongly influenced by meltwater characteristics (Leventer et al. 1982). Leventer et al. (1982) provided evidence for meltwater spikes in the Gulf of Mexico between about 16 000 and 14 000 years ago. There is also ample evidence for changes in ocean currents, sea-ice cover and abundance of foramanifera at this time (cf. Lehman and Keigwin 1992). Finally, it might well be that interpretation of the isotopic composition of ice from Greenland cores is made more complicated by the probability that the composition of the major moisture source — the surface water of the North Atlantic — may have deviated considerably from that of standard sea water. Indeed, floods may well explain some of the perplexingly abrupt changes in isotopic ratios in the late Wisconsinan parts of the cores.

Sea level, too, must have responded quickly if there were meltwater floods of the magnitude suggested. Blanchon and Shaw (1995) note that the coral *Acropora palmata*, which is highly sensitive to water depth, was submerged suddenly and drowned three times during the last deglacial phase. Wave-cut notches and coral platforms were also abandoned as sea level rose abruptly. Within the dating resolution available, the drowning events were synchronous with the major outburst floods during deglaciation. There should be evidence elsewhere of extreme rates of sea-level rise.

DISCUSSION AND CONCLUSIONS

A variety of evidence points to enormous outburst floods from the Laurentide ice sheet. Individual subglacial erosional and depositional landforms relate closely to presumed coherent structures in the flow that formed them and are best interpreted as meltwater forms. Absence of striation or poorly developed striation, percussion marks inferred to result from the collapse of cavitation bubbles, form similarity with landforms confidently ascribed to fluvial or eolian formation, internal structure (for depositional forms), and similar morphological setting to forms produced by wind and water in nature and in flumes all support the meltwater interpretation. Many of these characteristics contradict the more commonly held view that direct glacial processes, abrasion, subglacial deformation, or lodgement, produced most subglacial landforms.

Erosional marks in bedrock offer the most compelling evidence for the meltwater interpretation. They show a clear distinction between erosional scours (spindle flutes, sichelwannen and muschelbrüche) and remnant ridges (Beverleys and rat tails). Erosional scours are thought to begin with vortex impingement on the bed and to evolve to self-perpetuating forms that actually generate the eroding vortices. Remnant ridges are related to horseshoe vortices generated by bluff obstacles to flow.

Drumlins are also well-explained using this distinction between erosional scours and remnant ridges, although this explanation is more complicated than for erosional marks in bedrock. Beverleys are erosional remnants sculpted from bedrock or surficial sediment. Horseshoe vortices wrapped around obstacles are considered to have created hairpin scours in the initial stages of Beverly formation. With continued erosion, more streamlined Beverleys evolved in the form of classical drumlins with tapered tails. Livingstones are quite different in shape to Beverleys and resemble positive counterparts of erosional scours (spindle, parabolic and transverse asymmetrical). They are accretional and are interpreted to have been deposited in inverted erosional marks cut upwards into the ice bed by subglacial meltwater. Rogens are similarly interpreted. Some other drumlins with the same shapes as Livingstones may be erosional.

Flowlines drawn for regionally extensive fields of drumlins, flutings, Rogen moraine and bedrock erosional marks show coherent flow patterns over vast areas. Flow directions were remarkably consistent and unidirectional, and bedform size and shape changes gradually within fields. These observations point to the simultaneous formation of these subglacial meltwater forms on a regional scale. It follows that the scale of the floods thought to have formed these erosional marks must have been enormous. Conservative estimates for flow discharges forming subsets of drumlin and s-form fields are about 10^7 m^3/s, the total volume of flow for the Livingstone Lake field is estimated at about 8·4 $\times$ 10^4 km^3.

Two flow paths, the Livingstone Lake and Algonquin events (Figure 7.39), may be traced from close to the ice-sheet centre to close to its margin. There is no indication along these paths of ice-marginal deposition or erosion that suggest piecemeal formation of the landforms from which the flowlines are derived. Rather, later flows cross-cut the major paths, indicating the probability that the major routes involved a single event which took place when the ice sheet was close to its maximum extent. This evidence and evidence from Finland on the interaction between fan-shaped, drumlin-forming flows are most simply interpreted to mean that many subglacial meltwater forms were created simultaneously in a meltwater flow on the scale of the continental ice sheets. Deeply scoured basins, the Finger Lakes and Lake Mistassini and perhaps the Great Lakes themselves, may have been formed by such floods.

This interpretation greatly increases the magnitude of the proposed outburst flood events and begs the question of the meltwater source and the mechanism by which it was released. I present a conjectural model of how water may have been stored and released to produce outburst floods of the required magnitude (Figure 7.45). This model may be helpful as a guide for future analysis based on the physics it implies. Stagnant ice and a supraglacial lake, at first unconnected to the subglacial meltwater system and subsequently connecting suddenly, explain the water storage and provide sufficient potential energy for flow to overtop the outlets. In the specific case of the Laurentide ice sheet it is important that the outlet via Hudson Strait was sealed at this stage. This model also explains the cross-cutting relationships on the Western Plains and in the Lake Ontario Basin (Figure 7.39). Cross-cutting is inferred to result from late-stage flow diversion when the head of water in the supraglacial lake was insufficient to overtop the western outlets and to drive flow through the Finger Lakes troughs. Finally, this model goes a long way towards explaining why most glacial erosion and deposition occurred prior to drumlin and fluting formation and very little occurred afterwards. This asymmetry of glacial activity around drumlin-forming events is of utmost significance to ice-sheet modelling and inferences about deglaciation.

Ice over Hudson Strait may have collapsed as a consequence of back drainage into the depression formerly occupied by the supraglacial lake and of drawdown caused by the rise in sea level brought about by the outburst flood. In this case, ice from Keewatin and Labrador and water from the remnant of the supraglacial lake could drain through the Strait. Catastrophic drainage of this ice and water may account for a Heinrich layer at about 14 000 years ago. The reversal of drainage towards Hudson Strait may be an alternative to the traditional view of longstanding ice divides over Keewatin and Labrador/Ungava.

The meltwater floods may explain erosion in river valleys beyond ice margins and coarse-grained sediment in the Mississippi submarine fan and on the Hatteras abyssal plain. Enormous volumes of turbid meltwater are also expected to have had dramatic effects on the stratification and circulation of the North Atlantic Ocean. The possibility of these effects should be considered in reconstructions of lateglacial climate and in the interpretation of the oxygen isotopic composition of Greenland ice cores. Finally, the large outburst from the Laurentide ice sheet is expected to have caused an abrupt rise in sea level.

In summary, the erosional and depositional landforms of Figure 7.3 are confidently interpreted to be of meltwater origin. If the broader landscape interpretation is correct there must have been regional-scale outburst floods. The extents of proposed flood paths are well documented and presented with confidence. By contrast, the idea of simultaneous outburst flooding beneath a large part of the Laurentide ice sheet is new and is presented tentatively. So, too, is the suggestion that outbursts resulted from the catastrophic drainage of a large supraglacial lake. Nevertheless, even with regional-scale floods, the extraglacial effects of outburst floods are likely to have occurred. Naturally, the new ideas on flood magnitude are important regarding the degree of these effects.

ACKNOWLEDGEMENTS

Many people contributed to this work, especially my students and co-workers: Paul Blanchon, Tracy Brennand, Tim Fisher, Bob Gilbert, Phil Kor, Don Kvill, Lyn Murray, Bruce Rains, Dave Sharpe, Darren Sjogren and Bob Skoye. I am grateful for their support. Lee Weissling made many helpful comments and worked on the illustrations together with David Epp, Michael Fisher, Geoff Lester and Randy Pakan. Finally, I thank Derek Ford for encouragement.

REFERENCES

Alexander, H.S. 1932. Pothole erosion. *Journal of Geology*, **40**, 305–337.

Allen, J.R.L. 1971. Transverse erosional marks of mud and rock: their physical basis and geological significance. *Sedimentary Geology*, **5**, 167–385.

Allen, J.R.L. 1982. *Sedimentary Structures. Developments in Sedimentology* 30B. Elsevier, Amsterdam, vol. 2, 663 pp.

Alley, R.B., Blankenship, D.D., Bentley, C.R. and Rooney, S.T. 1987. Till beneath ice-stream B, 3, till deformation: Evidence and implications. *Journal of Geophysical Research*, **92**, 8921–8929.

Ashton, G.D. and Kennedy, J.F. 1972. Ripples on the underside of river ice covers. *ASCE Journal of the Hydraulics Division*, **98**, 1603–1624.

Aylsworth, J.M. and Shilts, W.W. 1989. Glacial features around the Keewatin ice divide, Districts of Mackenzie and Keewatin. *Geological Survey of Canada Paper* 88-24, 21 pp.

Baker, V.R. 1978. Large-scale erosional and depositional features of the Channeled Scabland. In Baker, V.R. and Nummedal, D. (eds.), *The Channeled Scabland*. National Aeronautical and Space Administration, Washington, DC, pp. 81–115.

Blanchon, P. and Shaw, J. Reef-drowning events during the last deglaciation: evidence for catastrophic sea-level rise and ice-sheet collapse *Geology*, **23**, 4–8.

Bond, G., Heinrich, H., Broecker, W., Labeyrie, L., McManus J., Andrews, J., Huon, S., Jantschik, R., Clasen, S., Simet, C., Tedesco, K., Klas, M., Bonani, G. and Ivy, S. 1992. Evidence for massive discharges of icebergs into the North Atlantic during the last glacial period. *Nature*, **360**, 245–249.

Bouchard, M.A. 1989. Subglacial landforms and deposits in central and northern Quebec, Canada, with emphasis on Rogen moraines. *Sedimentary Geology*, **62**, 293–308.

Boulton, G.S. 1974. Processes and patterns of glacial erosion. In Coates, D.R. (ed.), *Glacial Geology*. State University of New York, Binghampton, pp. 41–87.

Boulton, G.S. 1979. Processes of glacier erosion on different substrata. *Journal of Glaciology*, **23**, 15–37.

Boulton, G.S. 1987. A theory of drumlin formation by subglacial sediment deformation. In Menzies, J. and Rose, J. (eds), *Drumlin Symposium*. Balkema, Rotterdam, pp. 25–81.

Boulton, G.S. and Clark, C.D. 1990. The Laurentide ice sheet through the last glacial cycle: the topology of drift lineations as a key to the dynamic behaviour of former ice sheets. *Transactions of the Royal Society of Edinburgh: Earth Sciences*, **81**, 327–347.

Boulton, G.S., Smith, G.D., Jones, A.S. and Newsome, J. 1985. Glacial geology and glaciology of the last mid-latitude ice sheets. *Journal of the Geological Society of London*, **142**, 447–474.

Boyce, J.I. and Eyles, N. 1991. Drumlins carved by deforming till streams below the Laurentide ice sheet. *Geology*, **19**, 787–790.

Brennand, T.A. 1993. Laurentide meltwater systems: geomorphic and sedimentary evidence. Ph.D. thesis, University of Alberta, 292 pp.

Brennand, T.A. and Shaw, J. 1994. Tunnel channels and associated landforms, south-central Ontario: their implications for ice-sheet hydrology. *Canadian Journal of Earth Sciences*, **31**, 505–522.

Chapman, L.J. and Putnam, D.F. 1966. *The Physiography of Southern Ontario*, 2nd edition. University of Toronto Press, 386 p.

Christian, K.W. 1988. The Munro esker complex ice-contact sedimentation within a bedrock valley. M.Sc. thesis, Queen's University, Ontario, 216 pp.

Clark, C.D. 1993. Mega-scale glacial lineations and cross-cutting ice-flow landforms. *Earth Surface Processes and Landforms*, **18**, 18–29.

Dahl, R. 1965. Plastically sculptured detail forms on rock surfaces in northern Nordland, Norway. *Geografiska Annaler, Series A*, **47**, 3–140.

Dargahi, B. 1990. Controlling mechanism of local scouring. *Journal of Hydraulic Engineering*, **116**, 1197–1214.

DiLabio, R.N.W. 1981. Glacial dispersal of rocks and minerals at the south end of Lac Mistassini, Quebec, with special reference to the Icon dispersal train. *Geological Survey of Canada Bulletin*, **323**, 46 pp.

Dyke, A.S. and Morris, T.M. 1988. Drumlin fields, dispersal trains, and ice-streams in Arctic Canada. *Canadian Geographer*, **32**, 86–90.

Dyke, A.S. and Prest, V.K. 1987. Paleogeography of northern North America 18 000–12 000 years ago. *Geological Survey of Canada*, Map 1703, Sheet 1.

Dyson, J.L. 1952. Ice-ridged moraines and their relation to glaciers. *American Journal of Science*, **350**, 204–212.

Dzulynski, S. and Walton, E.K. 1965. *Sedimentary Features of Flysch and Greywackes. Developments in Sedimentology, Vol. 7*. Elsevier, Amsterdam, 274 pp.

Echelmeyer, K. and Harrison, W.D. 1990. Jakobshavns Isbrae, West Greenland: seasonal variations in velocity — or lack thereof. *Journal of Glaciology*, **36**, 82–88.

Fisher, T.G and Shaw, J. 1992. A depositional model for Rogen moraine, with examples from the Avalon Peninsula, Newfoundland. *Canadian Journal of Earth Sciences*, **29**, 669–686.

Fisk, H.N. 1944. *Geological Investigation of the Alluvial Valley of the Lower Mississippi River*. Mississippi River Commission, Vickburg, 78 pp.

Flint, R.F., Colton, R.B., Goldthwait, R.P. and Willman, H.B. 1959. *Glacial Map of the United States East of the Rocky Mountains*. Geological Society of America, scale 1:1 750 000.

Ford, D. and Williams, P. 1989. *Karst Geomorphology and Hydrology*. Unwin Hyman, London, 601 pp.

Fyfe, G.J. 1990. The effect of water depth on ice-proximal glaciolacustrine sedimentation: Salpausselkä I, southern Finland. *Boreas*, **19**, 147–164.

Gagnon, L. 1988. Géologie et geomorphologie de la Colline Blanche, région de Téiscamie, Québec. M.Sc. thesis, Université de Montréal, Canada, 148 pp.

Gilbert, R. and Shaw, J. 1994. Inferred subglacial meltwater origin of lakes on the southern border of the Canadian Shield. *Canadian Journal of Earth Sciences*, **31**, 1630–1637.

Gjessing, J. 1965. On 'plastic scouring' and 'subglacial erosion'. *Norsk Geogrask Tiddskrift*, **20**, 1–37.

Goldthwait, R.P. 1979. Giant grooves made by concentrated basal ice streams. *Journal of Glaciology*, **23**, 297–307.

Gorrell, G. and Shaw, J. 1991. Deposition in an esker, bead and fan complex. *Sedimentary Geology*, **72**, 285–314.

Gravenor, C.P. 1957. Surficial geology of the Lake Simcoe Area, Ontario. *Geological Survey of Canada Memoir*, **355**, 201 pp.

Greeley, R. and Iverson, J.D. 1985. *Wind as a Geological Process*. Cambridge University Press, 333 pp.

Hall, J. 1815. On the revolutions of the earth's surface. *Transactions of the Royal Society of Edinburgh*, **7**, 169–212.

Hammitt, F.G. 1980. *Cavitation and Multiphase Flow Phenomena*. McGraw-Hill, New York, 423 pp.

Hanvey, P.M. 1988. Sedimentology and genesis of Late Pleistocene drumlins in counties Mayo and Donegal, western Ireland. D.Phil. thesis, University of Ulster, Northern Ireland.

Heinrich, H. 1988. Origin and consequences of cyclic ice rafting in the northeast Atlantic ocean during the past 133,000 years. *Quaternary Research*, **29**, 142–152.

Hicock, S.R. 1988. Calcareous till facies north of Lake Superior, Ontario: implications for Laurentide ice streaming. *Géographie physique et Quaternaire*, **42**, 120–135.

Karcz, I. 1973. Reflections on the origin of source small-scale longitudinal streambed scours. In Morisawa, M. (ed.), *Fluvial Geomorphology*, State University of New York, Binghampton, pp. 149–173.

Kor, P.S.G., Shaw, J. and Sharpe, D.R. 1991. Erosion of bedrock by subglacial meltwater, Georgian Bay, Ontario: a regional view. *Canadian Journal of Earth Sciences*, **28**, 623–642.

Laverdière, C. and Guimont, P. 1980. Le vocabulaire de la géomorphologie glaciaire, IX. Terminologie illustrée des formes mineures d'érosion glaciaire. *Géographie physique et Quaternaire*, **35**, 363–377.

Laverdière, C, Guimont, P. and Dionne, J-C. 1985. Les formes et les marques de l'érosion glaciaire du plancher rocheux; signification, terminologie, illustration. *Paleogeography, Palaeoclimatology, Palaeogeography*, **51**, 365–387.

Lehman, S.J. and Keigwin, L.D. 1992. Sudden changes in North Atlantic circulation during the last glaciation. *Nature*, **356**, 757–762.

Leventer, A., Williams, D.F. and Kennet, J.P. 1982. Dynamics of the Laurentide ice sheet during the last glaciation: evidence from the Gulf of Mexico. *Earth and Planetary Science Letters*, **59**, 11–17.

Ljungner, E. 1930. Spaltektonik und morphologie der schwedishen Skagerakk-Kuste. Teil III. Die erosienformen. *Bulletin of the Geological Institutions of the University of Uppsala*, **21**, 255–475.

Lundqvist, J. 1969. Problems of the so-called Rogen moraine. *Sveriges Geologiska Undersöning*, **C. 648**, 32 pp.

MacAyeal, D.R. 1993. Growth/purge oscillations of the Laurentide ice sheet as a cause of the North Atlantic's Heinrich events. *Paleoceanography*, **8**, 775–784.

McCauley, J.F., Grolier, M.J. and Breed, C.S. 1977. *Yardangs of Peru and other desert regions*.

United States Geological Survey, Interagency Report, Astrogeology, **81**, 177 pp.
Mathews, E.B. 1917. Submerged "deeps" in the Susquehanna river. *Geological Society of America Bulletin*, **28**, 255–262.
Maxson, J.H. 1940. Fluting and faceting of rock fragments. *Journal of Geology*, **48**, 717–744.
Mellor, M. and McKinnon, G. 1960. The Amery Ice Shelf and its hinterland. *Polar Record*, **10**, 30–34.
Menzies, J. 1989. Towards a general hypothesis on the formation of drumlins. In Menzies, J. and Rose, J. (eds), *Drumlin Symposium*. Balkema, Rotterdam, pp. 9–24.
Menzies, J. and Rose, J. 1989. Subglacial bedforms — an introduction. *Sedimentary Geology*, **62**, 117–122.
Mollard, J.D. and Janes, J.R. 1983. *Airphoto Interpretation and the Canadian Landscape*. Energy Mines and Resources Canada, Ottawa, 415 pp.
Mullins, H.T. and Hinchey, E.J. 1989. Erosion and infill of New York Finger Lakes: implications for Laurentide ice sheet deglaciation. *Geology*, **17**, 622–625.
Mullins, H.T., Hinchey, E.J. and Muller, E.H. 1989. Origin of the New York Finger Lakes: a historical perspective on the ice erosion debate. *Northeastern Geology*, **11**, 166–181.
Murray, E.A. 1988. Subglacial erosional marks in the Kingston, Ontario, Canada, region: their distribution, form and genesis. M.Sc. thesis, Queen's University, Canada, 171 pp.
Pilkey, O.H. 1988. Basin plains; giant sedimentation events. In Clifton, H.E. (ed.), Sedimentologic Consequences of Convulsive Geological Events. Geological Society of America, *Special Paper*, **229**, pp. 93–99.
Prest, V.K. 1968. *Glacial map of Canada*. Geological Survey of Canada map 1253A, scale 1:5 000 000.
Prest, V.K. 1983. Canada's heritage of glacial features. *Geological Survey of Canada, Miscellaneous Report* 28, 119 pp.
Punkari, M. 1982. Glacial geomorphology and dynamics in the eastern parts of the Baltic Shield interpreted using Landsat imagery. *Photogrammetric Journal of Finland*, **9**, 77–93.
Rains, R.B., Shaw, J., Skoye, R., Sjogren, D. and Kvill, D. 1993. Late Wisconsin subglacial megaflood paths in Alberta. *Geology*, **21**, 323–326.
Rice, R.J. 1977. *Fundamentals of Geomorphology*. Longman, London, 387 pp.
Röthlisberger, H. 1972. Water pressure in intra- and subglacial channels. *Journal of Glaciology*, **11**, 177–203.
Russell, A.J. 1993. Supraglacial lake drainage near Sondre Stromfjord, Greenland. *Journal of Glaciology*, **39**, 431–433.
Selby, M.J. 1985. *Earth's Changing Surface*. Clarendon, Oxford, 607 pp.
Sharpe, D.R. and Shaw, J. 1989. Erosion of bedrock by subglacial meltwater, Cantley, Quebec. *Geological Society of America Bulletin*, **101**, 1011–1020.
Shaw, J. 1982. Melt-out till in the Edmonton area, Alberta, Canada. *Canadian Journal of Earth Sciences*, **19**, 1548–1569.
Shaw, J. 1983. Drumlin formation related to inverted melt-water erosional marks. *Journal of Glaciology*, **29**, 461–479.
Shaw, J. 1987. Glacial sedimentary processes and environmental reconstruction based on lithofacies. *Sedimentology*, **34**, 103–116.
Shaw, J. 1988a. Subglacial erosional marks, Wilton Creek, Ontario. *Canadian Journal of Earth Sciences*, **25**, 1256–1267.
Shaw, J. 1988b. Nothing new, nothing new. *Geoscience Canada*, **15**, 291–292.
Shaw, J. 1994. Hairpin erosional marks, horseshoe vortices and subglacial erosion. *Sedimentary Geology*, **91**, 269–283.
Shaw, J. and Gilbert, R. 1990. Evidence for large-scale subglacial meltwater flood events in southern Ontario and northern New York State. *Geology*, **18**, 1169–1172.
Shaw, J. and Kvill, D. 1984. A glaciofluvial origin for drumlins of the Livingstone Lake area, Saskatchewan. *Canadian Journal of Earth Sciences*, **21**, 1442–1459.
Shaw, J. and Sharpe, D.R. 1987. Drumlin formation by subglacial meltwater erosion. *Canadian Journal of Earth Sciences*, **24**, 2316–2322.
Shaw, J., Kvill, D. and Rains, R.B. 1989. Drumlins and catastrophic subglacial floods. *Sedimentary*

Geology, **62**, 177–202.
Shoemaker, E.M. 1992. Water sheet outburst floods from the Laurentide Ice Sheet. *Canadian Journal of Earth Sciences*, **29**, 1250–1264.
Smith, H.T.U. 1948. Giant glacial grooves in northwest Canada. *American Journal of Science*, **246**, 503–514.
Straw, A. 1968. Late Pleistocene glacial erosion along the Niagara Escarpment of Southern Ontario. *Geological Society of America Bulletin*, **79**, 889–910.
Thompson, G.H. 1985. The preglacial great falls of the Susqhehanna River. *Geological Society of America Abstracts with Programs*, **19**, 62.
Tinkler, R.J. and Stenson, R.E. 1992. Sculpted bedrock forms along the Niagara Escarpment, Niagara Peninsula, Ontario. *Géographie physique et Quaternaire*, **46**, 195–207.
Veillette , J.J. 1986. Former ice flows in the Abitibi-Timiskaming region: implications for the configuration of the late Wisconsinan ice sheet. *Canadian Journal of Earth Sciences*, **23**, 1724–1741.
Wakarani, N. 1993. Three dimensional, time-dependent flow over an upstream facing step. M.Sc. thesis, Queen's University, Canada, 175 pp.
Walder, J.S. 1982. Stability of sheet flow of water beneath temperate glaciers and implications for glacier surglng. *Journal of Glaciology*, **28**, 273–293.
Walker, R.G. 1992. Turbidites and submarine fans. In Walker, R.G. and James, N.P. (eds), *Facies Models*. Geological Association of Canada, pp. 239–263.
Weimer, P. 1989. Sequence stratigraphy of the Mississippi Fan (Plio-Pleistocene), Gulf of Mexico. *Geo-Marine Letters*, **9**, 185–272.
Wingfield, R. 1990. The origin of major incursions within the Pleistocene deposits of the North Sea. *Marine Geology*, **91**, 31–52.

Index